The BRICS Imperative

Ndivhuho Tshikovhi · Fulufhelo Netswera ·
Andréa Santos · Xiaolong Zou
Editors

The BRICS Imperative

Confronting Climate Challenges and Advancing Sustainable Development

palgrave
macmillan

Editors
Ndivhuho Tshikovhi
Faculty of Management Sciences
Durban University of Technology
Durban, South Africa

Fulufhelo Netswera
Research, Innovation
and Postgraduate Studies
University of Venda
Thohoyandou, South Africa

Andréa Santos
Transpot Engineering
Program
Federal University of Rio de Janeiro
Rio de Janeiro, Brazil

Xiaolong Zou
International and Public Affairs
Jilin University
Changchun, China

ISBN 978-3-032-05931-4 ISBN 978-3-032-05932-1 (eBook)
https://doi.org/10.1007/978-3-032-05932-1

Cover illustration: kenkuza_shutterstock.com

This Palgrave Macmillan imprint is published by the registered company Springer Nature Switzerland AG
The registered company address is: Gewerbestrasse 11, 6330 Cham, Switzerland

Preface

The BRICS Imperative: Confronting Climate Challenges and Advancing Sustainable Development brings together a rich collection of interdisciplinary scholarship that explores how the BRICS countries Brazil, Russia, India, China, and South Africa are responding to the urgent global challenges posed by climate change and the pursuit of sustainable development. The idea for this volume was born from the growing understanding that developing economies are not only significantly affected by the consequences of climate change but are also central to identifying and driving forward innovative, sustainable solutions. The BRICS nations represent diverse social, environmental, and economic contexts, yet share a collective responsibility and opportunity to shape more resilient and equitable futures. This book provides a space to reflect on these shared responsibilities through critical research, regional case studies, and thematic insights.

The contributions in this volume reflect a wide range of disciplines, including environmental studies, economics, public policy, marketing, education, agriculture, and energy. Together, these chapters offer a multilayered perspective on the intersection of climate action, economic resilience, and social justice within the BRICS context. The authors examine challenges ranging from food and water insecurity, energy transitions, and sustainable finance, to the complexities of consumer behaviour, green entrepreneurship, and disaster risk reduction. The book has been organized thematically to create a logical and coherent flow of ideas. The first section focuses on land use, agricultural practices, and food systems. This is followed by chapters that explore policy perspectives, energy cooperation, and the role of financial instruments. The later chapters move into sustainable marketing, behavioural change, and circular economy strategies. In doing so, the book offers not only critical reflection but also practical recommendations for enhancing cooperation and advancing the Sustainable Development Goals (SDGs), particularly in the Global South.

South Africa features prominently throughout this collection not only as a BRICS member but also as a nation that exemplifies many of the structural and environmental challenges facing the broader developing world. From rural livelihoods and community-based resilience to national policy alignment and innovation, South Africa's experience offers a useful lens for comparative and applied learning. It is our hope that this book will serve as a valuable resource for scholars, students, policymakers, and development practitioners. We believe the insights presented here will contribute to ongoing efforts to build climate-resilient systems, encourage cross-border learning, and advance socially just, inclusive development pathways across the BRICS and beyond.

Durban, South Africa — Ndivhuho Tshikovhi
Thohoyandou, South Africa — Fulufhelo Netswera
Rio de Janeiro, Brazil — Andrea Souza Santos
Changchun, China — Xiaolong Zou

Acknowledgements

We extend our sincere thanks to all contributing authors for their thoughtful scholarship, and to the peer reviewers for their constructive feedback which helped strengthen this volume. This book reflects the collaborative vision of the BRICS Research Institute, made possible through the support of the National Institute for the Humanities and Social Sciences (NIHSS). We are also grateful to the administrative team who assisted throughout the process. Finally, we thank the wider academic community for engaging with this work and for advancing ongoing conversations on climate action and sustainable development.

Introduction: Setting the Context

The climate crisis presents both a profound challenge and a catalytic opportunity for rethinking development pathways in the twenty-first century. While the impacts of climate change are global, their intensity and consequences are unequally distributed, with countries in the Global South particularly those in the BRICS bloc facing disproportionate vulnerabilities. This volume seeks to critically engage with this reality by exploring how the BRICS nations are responding to climate change through policies, innovations, and practices aligned with the principles of sustainable development. Rather than viewing climate change solely through the lens of environmental degradation, this book treats it as a multidimensional phenomenon one that is deeply intertwined with structural inequalities, economic resilience, political capacity, and social justice. In doing so, it recognizes that any meaningful response must be systemic, inclusive, and informed by context.

The BRICS countries offer a compelling case for study. They are home to over three billion people, possess vast ecological and economic diversity, and occupy pivotal roles in global governance debates. Their developmental models reflect a complex negotiation between industrialization and sustainability, growth and equity, sovereignty and multilateralism. These tensions are not barriers, but entry points for reimagining climate action that is both just and effective. The aim of this book is to examine climate-responsive strategies across key sectors and thematic areas in the BRICS context. Each chapter brings a unique disciplinary lens to bear ranging from agricultural sustainability, disaster recovery, and energy transitions to marketing, education, and circular economies. What binds these diverse contributions is a shared commitment to generating practical knowledge that supports the pursuit of low-carbon development.

Several themes emerge across the chapters:

- The need for context-specific climate strategies that reflect local socioeconomic realities.
- The importance of policy coherence and institutional capacity to support long-term climate planning.
- The role of knowledge production and education in shaping sustainability practices.
- The growing relevance of financial innovation and new business models to drive green transitions.
- And the imperative to close the gap between climate awareness and behaviour change, especially in consumer markets.

The structure of the book reflects an intentional logic, beginning with land use and agricultural adaptation, moving through socio-economic and policy frameworks, and culminating in strategies for communication, consumption, and circularity. While each chapter stands alone in its analytical rigour, collectively they offer a layered understanding of how BRICS nations are navigating climate change within their development trajectories. As climate challenges become more frequent and complex, there is an urgent need for scholarship that is not only critical but also constructive highlighting what works, what is possible, and what is needed.

Overview of the Book

This book is composed of twelve chapters, each of which contributes a critical perspective to the broader understanding of climate action and sustainable development within the BRICS countries. The chapters have been carefully arranged to create a logical progression from environmental management and food security to policy frameworks, financial instruments, and behavioural change. The first three chapters focus on land use, agriculture, and food systems, highlighting the urgent need for climate-compatible practices that build resilience and support food security. Chapter 1 begins with an in-depth systematic review of agricultural waste management strategies in South Africa, offering evidence of how circular innovations can contribute to both environmental and economic sustainability. Chapter 2 turns attention to indigenous leafy vegetables as a pathway for poverty alleviation and nutrition security in rural South African communities. Chapter 3 introduces agrivoltaic systems as a promising solution that integrates sustainable energy generation with agricultural productivity across BRICS countries.

Chapter 4 shifts the lens towards education and knowledge systems. It examines how Brazil and South Africa have responded to the Sustainable Development Goal 4 through education for sustainable development, identifying both achievements and persistent challenges in building a more climate-literate society. Chapters 5 through 7 examine policy and institutional dimensions. Chapter 5 provides a comparative policy analysis of climate and development frameworks across BRICS, with a particular emphasis on South Africa's role and alignment within the group. Chapter 6 explores the socio-economic impacts of climate-induced disasters, with the 2023 KwaZulu-Natal floods as a focal case, and argues for improved integration between disaster response and long-term climate resilience planning. Chapter 7 investigates the transformative role of emerging financial instruments such as green bonds and sustainability-linked loans in advancing sustainable infrastructure development amid climate change uncertainties.

Chapter 8 continues the discussion on institutional cooperation by looking at BRICS energy cooperation, especially how the member states are responding to global energy transitions and supply threats. The chapter highlights new models of intergovernmental and financial coordination to promote energy security while shifting towards renewables. Chapter 9 builds on the previous discussions by addressing the broader trends in sustainable growth across the BRICS nations. It assesses the different pathways that member states are pursuing in climate mitigation, with particular attention to carbon leakage, taxation, and regional policy coherence.

The final three chapters explore the role of consumer behaviour, marketing, and circular economic practices. Chapter 10 presents a systematic review of sustainable marketing strategies in South Africa and assesses their effectiveness in shifting consumer habits towards responsible consumption. Chapter 11 investigates the relationship between marketing and carbon emissions reduction, highlighting the barriers to behavioural change in low-income and vulnerable communities. Chapter 12 concludes the volume by providing a systematic review of circular economy practices in BRICS countries. It discusses the policy opportunities and implementation challenges that hinder progress towards waste reduction, resource efficiency, and sustainable production systems. Together, these chapters provide an integrated view of how BRICS countries are tackling climate-related challenges. The book makes a timely contribution to both academic and policy debates by offering actionable insights, cross-country comparisons, and evidence-based reflections on the role of emerging economies in building a just and sustainable future.

Contents

Editors and Contributors

Mr. Abraham Adeniran is a Ph.D student in the Department of Public Administration and Economics at Durban University of Technology, South Africa, where his research focuses on the sustainability of solar energy projects through integrated monitoring and evaluation. He is a research and development professional with over eight years of experience, bringing expertize in energy economics, Sustainable Development Goals, digital transformation, data analytics, and project management. In his role at Afe Babalola University, Abraham supports the Office of the Vice Chancellor in coordinating SDG-aligned activities and interdisciplinary research initiatives to drive innovation and lasting impact.

Dr. Hintsa Araya is a Senior Researcher at the Agricultural Research Council. His research focus is on the indigenous/African Leafy Vegetables, water use efficiency, and plant nutrition.

Prof. Michael Bairu is a research team manager at the Agricultural Research Council's (ARC) Vegetable, Industrial, and Medicinal Plants Research Campus.

Mr. Ravi Kumar Chopdar is an Assistant Professor in the Department of Physics at Government Girls College, Bundi, Rajasthan, India. He is also a research scholar at the University of Kota, Rajasthan, specializing in the applications of agrivoltaic systems. With a keen focus on integrating sustainable energy solutions, Mr. Chopdar has presented over 10 research papers at various international seminars and conferences. He has also contributed to the academic community with publications in reputed journals.

Dr. Carlo Emile Daniels obtained his Ph.D. in 2025, specializing in Comparative and International Education from North-West University in South Africa. He currently holds the position of Senior Lecturer in the Department of Educational Foundations at the University of South Africa, where he instructs courses related to Comparative and International Education. His research interests are centred on Education for Sustainable Development and the enhancement of research capacity, particularly within the context of BRICS nations. Dr. Daniels is committed to advancing sustainable educational practices that empower communities and strengthen research capabilities across diverse educational settings.

Mr. Simphiwe Emmanuel Dube is a lecturer and Programme Coordinator in Tourism at Nelson Mandela University. Simphiwe Dube holds a master's degree in management sciences specializing in Tourism and Hospitality Management obtained at Durban University of Technology. He is also currently registered for his doctorate degree in Management Sciences specializing in Tourism and Hospitality Management at Durban University of Technology. Simphiwe's research interests include marketing for tourism, customer service, and tourism higher education curriculum development. He consulted in the Tourism Research studies for five years prior joining the academia. His research focuses on areas such as Tourism Development Strategies, Economic Impact Assessments, Event Impact Assessments, Feasibility Studies, and Market Research studies.

Ms. Tanishka Gupta is an Assistant Professor at Jagannath International Management School, Vasant Kunj, New Delhi, India, in the Department of Management. She holds a Master of Commerce degree from the Delhi School of Economics and has four years of teaching experience. Prior to joining academia, she worked as a Research Associate at the Indian Council of Social Science Research. Her research interests include Finance and International Business.

Mr. Manaka J. Makgato is a dedicated Soil and Crop Science Researcher focused on bolstering sustainable agriculture in South Africa. With strong academic credentials and a growing publication record, he plays a key role in advancing climate-resilient farming, indigenous crop utilization, and water-efficient practices for smallholders and marginal lands. His work at ARC underscores a commitment to addressing food insecurity through science-based innovation.

Dr. Salmina Mokgehle is a Researcher and Lecturer at the University of Mpumalanga, specializing in medicinal plants. Her work emphasizes sustainable production practices and the conservation of indigenous crop species. She promotes indigenous knowledge systems, biodiversity conservation, and inclusive innovation. Dr. Mokgehle has served as an associate editor for both the *South African Journal of Science* and the *South African Journal of Botany*. She is rated as an Y2 researcher by the National Research Foundation of South Africa. Her academic contributions span teaching, community engagement, and international research publications.

Ms. Basiiraa Moosa is a Lecturer at the University of Johannesburg, South Africa. Her research focuses on Climate Change, UN SDGS, and Climate Resilience. She is a young academic currently growing her research portfolio and seeking to make meaningful contribution to the discipline of Urban Planning.

Prof. Fhatuwani Nixwell Mudau is an Acting Deputy Vice-Chancellor and Head of the College of Agriculture, Engineering, and Science at the University of KwaZulu-Natal.

Dr. Azwimbavhi Reckson Mulidzi obtained his Ph.D. in Geology and in environmental management from Stellenbosch University, Master's degree in Land Use Planning from University of Pretoria. Dr. Mulidzi is a Research Team Manager for Soil and Water Science division at ARC Institute for Fruits, Vine, and Wine in Stellenbosch. His research interest includes: Conservation Agriculture, Water Resource management, and Management of natural resources. Dr Mulidzi is involved in lot of projects that deals with student development and mentoring. Dr Mulidzi has established and continues to maintain a strong relationship with Cape Peninsula University of Technology (CPUT), University of the Western Cape (UWC), University of Limpopo, and University of Pretoria wherein He is the co-supervisor of several post graduate students. Dr. Mulidzi is a qualified Professional Natural Scientist (Pr. Sci. Nat). Dr. Mulidzi has published more than 30 peer-reviewed papers and over 20 semi-scientific papers at national and international journals. In 2022, Dr Mulidzi was appointed as a member to serve on the committee on Science and Technology under the United Nations Convention to combat Desertification in South Africa. In 2024, Dr. Mulidzi was nominated by ARC CEO as a Project Counterpart on the Regional Project RAF5092, titled Enhancing Agricultural Productivity for Improved Food Security in Africa (AFRA)

Prof, Fulufhelo Netswera is a Deputy Vice Chancellor for Research and Postgraduate Studies at the University of Venda, South Africa.

Dr. Paballo Patience Ntobaki is a Lecturer and Researcher in Marketing and Retail at the Durban University of Technology. Her research explores digital transformation, sustainability, the circular economy (CE), and global value chains (GVCs) in marketing and retail.

Ms. Zikho Qwatekana is a dedicated Ecotourism Lecturer at the Durban University of Technology (DUT) in South Africa. With over a decade of academic experience, she has made significant contributions to teaching, curriculum development, and academic support in the fields of tourism and environmental management. She holds a Master of Management Sciences in Hospitality and Tourism and has recently submitted her Ph.D. in Management Sciences, with a focus on Ecotourism and Environmental Management, for examination. Her research interests include ecotourism, climate change resilience, rural development, tourism governance, and heritage management. Zikho has published in peer-reviewed journals, contributed to edited volumes, and presented her work at various international conferences. Her academic portfolio reflects a strong commitment to advancing inclusive and sustainable tourism practices, particularly within the Global South. She has also recently begun supervising postgraduate students, marking the start of her journey in academic mentorship and research leadership.

Dr. Willem Jansen van Rensburg is a Senior Researcher at the Agricultural Research Council.

Andréa Santos is the Executive Secretary of the Brazilian Panel on Climate Change and Executive Manager of Green Fund of the Federal University of Rio de Janeiro.

Mr. Filip Selamovski currently serves as an Assistant Professor at the Department of International Business of the Plekhanov Russian University of Economics where he conducts research focusing on international business challenges and cross-border economic cooperation. At the same time Selamovski is pursuing an advanced Ph.D. in global economic affairs. In addition to his academic pursuits, he co-founded a consulting firm focused on providing strategic advice to firms navigating complex international markets and sustainability issues.

Dr. Namrata Sengar is presently working as an Assistant Professor at Department of Pure and Applied Physics, University of Kota, Kota India. She completed her Ph.D. in solar energy utilization in 2009. She was awarded with research fellowships and best paper presentation awards twice. Four collaborative research projects with UK universities in the area of solar energy were successfully completed by her, and two projects are ongoing. She has published more than 25 research papers in journals and delivered more than 50 presentations in conferences/seminars.

Dr. Yogesh Sharma is an Assistant Professor at the University School of Management and Entrepreneurship (USME), Delhi Technological University, India. He holds a Ph.D. from Jawaharlal Nehru University, an MBA from Jamia Millia Islamia and BBS from University of Delhi. A UGC-Senior Research Fellow and triple NET qualifier, Dr. Sharma has academic and industry experience across institutions like University of Delhi, CII, FICCI, ASSOCHAM, NFCG and IFFCO. Internationally, he has been associated with the BRICS Research Institute in South Africa and has participated in academic programs in Brazil, Russia, China, and South Africa. He has published extensively in reputed journals indexed in ABDC, ABS, SCI, Scopus, and Web of Science. His research interests include Marketing, Subscription Models, Servitization, Green Marketing, Sustainability, AI, Data Analytics, Neural Networks, and BRICS related studies.

Ms. Svetlana Evgenyevna Sobenina is an economist specializing in global economics and data science. Her research focuses on international trade patterns, sustainable economy and economic forecasting of long-term trends in developing countries using advanced statistical models. In addition to her academic credentials, Sobenina has extensive corporate experience in project development, including IT and sustainability-related projects.

Dr. Ankit Suri is an Assistant Professor at Motilal Nehru College, University of Delhi. He holds a Ph.D. in Management from Jawaharlal Nehru University and a Master's in Commerce from the Delhi School of

Economics. His research interests include Financial Literacy and Inclusion, the Informal Sector Labour Market, and Sustainable Development. Dr. Suri has presented research papers at several prestigious international conferences, including those held in Brazil, Russia, China, South Africa, and at the World Bank Headquarters. He has multiple publications in reputed academic journals and remains actively involved in research, teaching, and contributing to policy-relevant discussions in his areas of expertize.

Ms. Aifani Confidence Tahulela is a researcher affiliated with the Durban University of Technology, South Africa. Their research focuses on circular economy, waste management, sanitation, and sustainable development, with an emphasis on service delivery, eco-innovation, and resource efficiency in emerging economies. Tahulela's work contributes to advancing environmental sustainability through the integration of policy, technology, and geographic information systems (GIS).

Dr. Mariette Truter is a Research Team Manager in Leguminous Leafy Vegetables. Dr. Truter is a plant pathologist specializing in Mycology at the Agricultural Research Council. Her current research interests include the characterization and identification of fungal phytopathogens of vegetables, medicinal, and ornamental plants.

Dr. Ndivhuho Tshikovhi is Research Fellow in the Faculty of Management Sciences at Durban University of Technology, South Africa.

Dr. Ksenia Voronina is a Lecturer at the Moscow State Institute of International Relations (MGIMO University). In 2024, she received a doctoral degree in Economics with a Ph.D. thesis titled *"Economic Aspects of the Transformation of the Russian Oil Industry in the Context of the Global Energy Transition."* In 2022, Ksenia completed a Master's degree in International Law, and in 2020, she received a Bachelor's degree in International Business and Business Administration, both at MGIMO University.

Ms. Thembeka Goodness Zondi is a Master's student in Marketing at Durban University of Technology. Her research focuses on sustainable marketing and consumer behaviour.

Xiaolong Zou is an Associate Professor at the School of International and Public Affairs at Jilin University and is also a Researcher at the Institute of National Development and Security Studies at the same university.

Dr. Nkululeko PraiseGod Zungu is a dedicated Senior Lecturer in the Department of Marketing and Retail Management, in the Faculty of Management Sciences, with a deep passion for academic and research. With a strong background in marketing and an extensive educational journey. His research interests span a wide range of marketing topics, including Service Quality, Social Media Marketing, Artificial Intelligence Marketing, Sports Sponsorship, Marketing Strategy, and Entrepreneurship Marketing. His commitment to research is underscored by his active role in supervising Master's and Doctoral students. He holds several qualifications including Ph.D. in Man. Sciences: Marketing (DUT), M.Tech. Degree in Marketing (DUT), PG Dip. in Man. Sciences: Bus. Admin (DUT), Adv Dip in Man. Sciences: Bus. Admin. (DUT) BTech Degree in Marketing (DUT), N. Dip. in Marketing (DUT), Teaching with Technology Certi. (UCT), and Strengthening PG Cert. (RU).

List of Figures

Chapter 1

Chapter 2

Chapter 3

Chapter 5

Chapter 7

Chapter 8

Chapter 9

Chapter 10

Chapter 1

Chapter 12

List of Tables

Chapter 7

Chapter 10

Chapter 11

CHAPTER 1

Climate-Compatible Agricultural Waste Management Strategies for Resilient Land Use in South Africa: A Systematic Review of Circular Innovations

Aifani Confidence Tahulela, Yogesh Sharma, and Ankit Suri

1 Introduction

Agricultural waste management (AWM) lies at the nexus of climate change mitigation, sustainable land use, and rural development. Globally, the agriculture sector contributes significantly to greenhouse gas (GHG) emissions, particularly methane (CH_4) and nitrous oxide (N_2O), arising from the mismanagement of organic waste such as manure, crop residues, and agro-industrial byproducts (IPCC, 2022). Without climate-aligned

A. C. Tahulela (✉)
Department of Public Management, Durban University of Technology, Steve Biko Campus, Durban, South Africa
e-mail: aifaniT@dut.ac.za

Northeastern University School/Faculty of Public Administration, Shenyang, People's Republic of China

N. Tshikovhi et al. (eds.), *The BRICS Imperative*,
https://doi.org/10.1007/978-3-032-05932-1_1

interventions, these emissions threaten to undermine global efforts to limit warming and achieve resilience in food production systems. The urgency is more pronounced in the Global South, where weak regulatory structures, limited infrastructure, and growing demographic pressures compound the environmental and socio-economic risks of unsustainable agricultural practices (Brenya et al., 2022a, 2022b; Miladinov, 2023). In South Africa, approximately 28 million tonnes of agricultural waste are generated annually (Ryan et al., 2014), yet the absence of a coordinated AWM framework represents both an ecological liability and a missed opportunity for climate-compatible development (FAO, 2022).

Current disposal methods, including open burning, landfilling, and unregulated spreading, not only accelerate land degradation but also contribute directly to the emissions of short-lived climate pollutants. These practices conflict with South Africa's commitments under its Nationally Determined Contributions (NDCs) to the Paris Agreement, the National Development Plan (NDP) 2030, and the recently promulgated Climate Change Bill. To address this dual challenge of emissions and degradation, a paradigmatic transition is required from linear, extractive production models to circular bioeconomy frameworks. These approaches valorize agricultural residues through biological processes such as composting, anaerobic digestion, thermochemical processes such as pyrolysis, and biotechnological processes such as fermentation that enhance nutrient recycling, generate renewable energy, and reduce dependence on fossil inputs (Brenya et al., 2022a, 2022b; Mahajan, 2025; Miladinov, 2023). Importantly, such strategies generate climate co-benefits by simultaneously improving soil health, lowering emissions, and increasing farm-level resilience to climate variability.

Y. Sharma
Delhi School of Management, Delhi Technological University, New Delhi, India

A. Suri
Motilal Nehru College, University of Delhi, New Delhi, India

This shift is theoretically anchored in systems thinking, life cycle sustainability assessment (LCSA), and the waste hierarchy, all of which emphasize resource efficiency and long-term environmental integrity. Emerging empirical research increasingly supports the viability of low-carbon waste-to-resource pathways. For instance, the application of animal-derived organic fertilizers significantly enhances soil carbon retention by increasing soil organic carbon stocks (Kamilaris et al., 2020). Meanwhile, biochar-enriched composting systems substantially reduce greenhouse gas emissions during composting, notably CH_4 by ~91% and N_2O by ~56% in cattle-manure treatments (Smith et al., 2023; Taghizadeh-Toosi et al., 2021). Similarly, biomass-based energy production offers a dual benefit of climate mitigation and improved energy access in low-income regions, despite substantial infrastructural and financial bottlenecks (IEA, 2024; Teklu et al., 2021).

Furthermore, vermitechnology, the use of earthworms to biologically convert biowaste into high-value compost, has gained traction as a climate-smart innovation that supports low-input, regenerative farming, especially among smallholders (Dada, 2024; Frontiers, 2022; Robatjazi, 2023). Despite these advances, several barriers persist: limited local validation of composting protocols, insufficient modeling of GHG abatement potential, and poor integration of waste management into national climate policy frameworks. Given these critical gaps, this study conducts a systematic literature review (SLR) to synthesize existing evidence on climate-compatible AWM strategies with a focus on South African agriculture. Guided by the PRISMA protocol and structured through bibliometric coupling, thematic clustering, and spatial synthesis, this review aims to:

- Categorize current AWM strategies along technological, ecological, and socio-economic dimensions.
- Assess their contribution to climate change mitigation, agroecosystem adaptation, and sustainable land management.
- Identify innovations with circular economic potential and climate co-benefits.
- Highlight key research and policy gaps to inform future climate-resilient waste management interventions.

By embedding the review within a climate policy framework, this study contributes to a growing body of scholarship advocating for integrated, low-carbon transitions in agricultural systems that are essential for realizing the ambitions of the Paris Agreement, the Sustainable Development Goals (SDGs), and just transitions in the Global South.

2 Materials and Methods

2.1 *Systematic Review Framework*

This review adhered to the Preferred Reporting Items for Systematic Reviews and Meta-Analyses PRISMA protocol, ensuring methodological transparency, reproducibility, and rigour. PRISMA's four-phase approach identification, screening, eligibility, and inclusion) was implemented to document the selection process and minimize bias. The study was designed to generate an integrative evidence base to inform climate-compatible agricultural waste management (AWM) strategies in the South African context.

To enhance analytical robustness, the review also incorporated principles of Life Cycle Thinking (LCT) and climate systems analysis, recognizing that sustainable AWM strategies must be evaluated not only for technical efficacy but also for their mitigation potential, adaptive capacity, and contribution to broader climate resilience.

2.2 *Search Strategy*

A comprehensive multi-source search was performed between January and March 2024, using Scopus as the primary database, supplemented by searches in Google Scholar, ScienceDirect, and Web of Science. Grey literature including policy reports and technical briefs from the FAO, UNEP, and South African government portals was also reviewed to capture context-specific insights. Search terms were structured using Boolean operators and truncated syntax, including:

> TITLE-ABS-KEY ("Waste Management Strategies" OR "Waste Management" OR "Waste Reduction" OR "Waste Disposal") AND TITLE-ABS-KEY ("Sustainable Agriculture" OR "Sustainable Land Use" OR "Agricultural Sustainability") AND)PUBYEAR > 2013 AND PUBYEAR <

2026 AND (LIMIT-TO (LANGUAGE, "English")) AND (LIMIT-TO (DOCTYPE, "ar") OR LIMIT-TO (DOCTYPE, "re") OR LIMIT-TO (DOCTYPE, "ch"))AND (LIMIT-TO (PUBSTAGE, "final")

The search was limited to peer-reviewed publications in English, published between 2013 and 2024. The initial SCOPUS query yielded 489 records. After the application of inclusion criteria including publication type, language, geographical relevance, and alignment with climate or sustainability outcomes 338 articles were retained for preliminary review. A second-level manual screening, based on thematic alignment and methodological quality, reduced the corpus to 114 articles. These were subjected to bibliographic coupling analysis, which resulted in the identification of 10 highly interconnected studies forming four thematic clusters central to the final synthesis.

This structured and iterative selection process is summarized in Fig. 1 (PRISMA flow diagram), and ensured that only studies with clear methodological rigour, policy relevance, and climate-smart innovation potential were included.

2.3 *Inclusion and Exclusion Criteria*

In order to maintain methodological transparency and thematic accuracy, we defined our inclusion and exclusion criteria beforehand. These parameters supported the systematic identification of studies relevant to **climate-smart agricultural waste management (AWM)** in the South African context, ensuring that only academically rigorous and thematically coherent sources informed the final synthesis. The applied screening logic is detailed in Table 1, which outlines the inclusion and exclusion criteria used during article selection.

2.4 *Data Extraction Procedure*

Data extraction was conducted systematically using a pre-defined template, which facilitated consistency and reproducibility across all reviewed studies. The extraction process was supported by NVivo 12 qualitative analysis software, enabling efficient data handling, coding, and thematic classification. The following data fields were systematically captured:

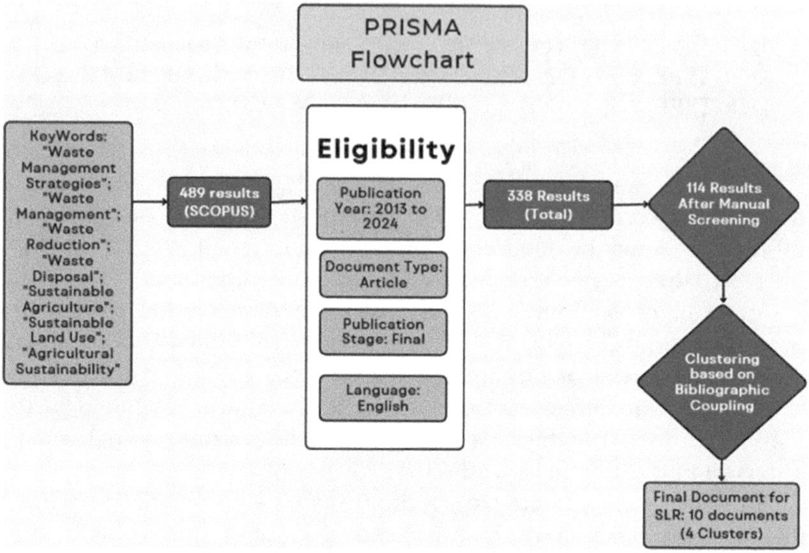

Fig. 1 PRISMA flow diagram showing study selection and inclusion

Table 1 Inclusion and exclusion criteria for study selection

Criteria category	*Inclusion criteria*	*Exclusion criteria*
Topical scope	Studies focused on *waste management* within *sustainable agriculture* or *land use* contexts	Studies unrelated to agriculture or dealing with non-agricultural waste (e.g., industrial, e-waste)
Publication year	2014–2025	Studies published before 2014 or future-dated beyond 2025
Language	English	Non-English publications
Document type	Peer-reviewed articles, reviews, and book chapters	Editorials, theses, commentaries, opinion pieces
Publication stage	Final publication status with full-text availability	In-press, abstract-only, or inaccessible sources
Geographical relevance	South African or contextually analogous agroecological regions	Studies with no relevance to Sub-Saharan agriculture or climate contexts
Methodological rigour	Empirical or conceptual studies with a structured design	Anecdotal, grey literature, or methodologically weak publications

- Type of Waste Management Strategy: Including composting, vermitechnology, biochar application, anaerobic digestion, and integrated manure systems.
- Climate-Related Outcomes: Explicit attention was given to indicators of climate mitigation (e.g., greenhouse gas reduction, carbon sequestration) and climate adaptation (e.g., improved soil moisture retention, drought resilience).
- Innovation Characteristics: Technologies and practices were examined for innovative elements, such as biochar integration, vermitechnology applications, or waste-to-energy transformations.
- Methodological Approaches: The review documented research methodologies such as life cycle assessments (LCA), field trials, meta-analyses, and modelling techniques, which underpin the robustness and generalizability of findings.
- Policy and Socio-Economic Relevance: Studies were assessed for their alignment with policy frameworks, economic viability, and implications for rural livelihoods and sustainable development goals (SDGs).

This structured approach ensured that both qualitative and quantitative dimensions were captured, and that contextual factors relevant to the South African agroecological environment were not overlooked.

2.5 *Thematic Synthesis*

The extracted data were analyzed using inductive thematic analysis, allowing for the emergence of patterns, relationships, and dominant discourses within the literature. Four principal thematic clusters were identified:

2.5.1 Bioenergy and Circular Bioeconomy

This cluster encapsulated technologies that convert waste into renewable energy (e.g., biogas, syngas), aligning with broader circular economy principles. These studies typically discussed resource valorization, energy recovery, and systemic sustainability benefits.

2.5.2 *Vermitechnology and Biowaste Bioconversion*

Research in this category emphasized the biological conversion of organic waste through earthworms, highlighting vermicomposting as a low-cost, decentralized solution with multiple ecological and agronomic co-benefits.

2.5.3 *Composting and Manure Management Systems*

This thematic area focused on traditional and enhanced composting methods, the management of livestock waste, and the integration of composted materials into sustainable land management practices.

Each thematic cluster was evaluated in relation to its:

- Contribution to Climate Change Mitigation, such as the reduction of methane and nitrous oxide emissions;
- Support for Climate Adaptation, including improvements in soil structure, water retention, and agroecosystem resilience;
- Degree of Systemic Innovation, especially decentralized and community-led waste-to-resource initiatives relevant to South African rural and peri-urban contexts.

2.6 *Synthesis Outcomes*

The synthesis revealed a growing body of context-specific innovations in AWM within South Africa, yet also highlighted gaps in policy integration, scalability, and interdisciplinary collaboration. While bioeconomy approaches and climate-smart technologies are gaining prominence, further empirical work is required to assess long-term impacts on ecosystem services and rural livelihoods. The thematic synthesis thus offers a comprehensive lens through which current practices can be assessed and future research trajectories can be proposed.

2.7 *Analytical Tools and Climate Relevance*

To examine intellectual coherence and thematic convergence among the selected studies, bibliographic coupling was conducted using VOSviewer version 1.6.18. This scientometric technique facilitated the visualization of citation linkages and enabled the identification of structural clusters corresponding to four thematic domains. In parallel, a GIS-based spatial analysis was performed using ArcGIS version 10.8 to assess regional disparities in research coverage. This approach highlighted geographic gaps across South Africa's diverse agroecological zones, where spatial heterogeneity driven by variations in soil properties, rainfall distribution, and farming systems plays a critical role in climate adaptation planning.

In addition to spatial and bibliometric analyses, the selected literature was systematically reviewed for the integration of climate-relevant analytical tools. These included Life Cycle Assessment (LCA) for evaluating environmental performance, Decision Support Tools (DSTs) to aid in strategic decision-making, greenhouse gas (GHG) quantification frameworks consistent with IPCC Tier I and Tier II methodologies, and techno-economic assessments for feasibility and scalability evaluation. To ensure the reliability of thematic coding, 20% of the final sample was independently double coded by two reviewers. The intercoder reliability test yielded a Cohen's Kappa coefficient of 0.82, indicating a high level of agreement and confirming the robustness of the content analysis.

3 Results

3.1 *Overview of Included Studies*

The final dataset consisted of ten peer-reviewed studies published between 2016 and 2022, selected through bibliographic coupling and thematic alignment with the principles of climate-smart agricultural waste management (AWM). While the studies originated from diverse geographic contexts, each was critically assessed for its relevance to the South African agroecological and policy landscape. The studies collectively addressed key dimensions of climate-responsive AWM, including the reduction of greenhouse gas (GHG) emissions, particularly methane and nitrous oxide, enhancement of soil health through organic matter restoration, renewable energy generation from agro-waste streams, and the integration of circular bioeconomy principles within farming systems.

These contributions stemmed from a range of interdisciplinary fields, including sustainability science, agroecology, environmental engineering, and systems thinking. Each article addressed at least one facet of climate mitigation, adaptation, or resource circularity, thereby aligning with both national frameworks such as South Africa's Nationally Determined Contributions (NDCs) and global commitments under Sustainable Development Goal 13 (Climate Action). A synthesis of the four thematic clusters, along with representative insights and sources, is presented in Table 2 to provide a consolidated view of the conceptual and empirical contributions across the reviewed literature. To complement this synthesis, Fig. 2 offers a visual representation of the bibliographic coupling network, where the proximity of nodes and the thickness of edges denote the degree of thematic similarity and citation connectivity among the selected studies. These visual and tabular syntheses reinforce the four emergent thematic clusters:

1. Organic Fertilization and Soil Health Enhancement.
2. Bioenergy and Circular Bioeconomy Valorization
3. Vermitechnology and Biowaste Bioconversion.
4. Composting and Manure Management Systems

Table 2 Thematic synthesis of waste management clusters

Cluster	*Thematic focus*	*Key insights*	*Representative studies*
1. Organic Fertilization & Soil Health	Composting, biochar integration, animal-derived fertilizers	Enhances microbial activity, reduces reliance on synthetic fertilizers, mitigates GHG emissions	Bhunia et al. (2021), Harindintwali et al. (2021), Zhang et al. (2021)
2. Bioenergy and Circular Bioeconomy	Biomass valorization, biofuel production	Converts agricultural residues into bioenergy and bioproducts; supports energy security and emissions reduction	Gupta et al. (2022), Karimi Alavijeh and Yaghmaei (2016), Koul et al. (2022)
3. Vermitechnology & Biowaste Management	Earthworm-assisted composting, nutrient cycling	Enhances soil fertility, diverts waste from landfills, supports low-input farming	Singh et al. (2020), Yuvaraj et al. (2021)
4. Manure Management Systems	On-farm composting, anaerobic digestion	Reduces odour, improves nutrient recovery, offers co-benefits like biogas and water reuse	Pergola et al. (2018), Varma et al. (2021)

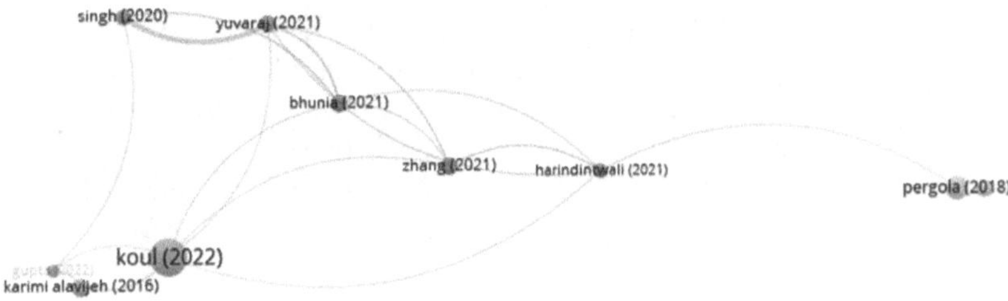

Fig. 2 Bibliographic coupling network of the reviewed studies, illustrating thematic similarity and co-citation linkages across the four identified clusters

Each cluster is elaborated in Sects. 3.2 to 3.5, where climate impacts, innovations, and implementation challenges are critically discussed.

3.2 *Thematic Cluster 1: Organic Fertilization and Soil Health Enhancement*

This thematic cluster centres on the substitution of synthetic fertilizers with organic amendments derived from agricultural residues, livestock manure, and biochar-enriched compost. These interventions are fundamental to climate-smart agriculture, offering a dual function: (1) GHG mitigation by reducing emissions associated with industrial fertilizer production and application, and (2) climate adaptation through the enhancement of soil structure, microbial activity, and water retention capacity, critical factors under increasing climatic variability.

3.2.1 *Key Climate and Environmental Insights*

Zhang et al. (2021) conducted a meta-analysis of 68 datasets and demonstrated that the optimization of composting parameters, particularly the carbon-to-nitrogen (C:N) ratio, moisture content, and aeration can reduce total nitrogen losses by up to 38% and greenhouse gas (GHG) emissions by as much as 40%. These findings underscore the critical role of process control innovations in enhancing the climate mitigation potential of organic amendments. In the broader context of circular and technological innovation, biochar-based composting has emerged as a frontier strategy that not only enhances nutrient-use efficiency but also enables long-term carbon sequestration in soils, positioning it as a pivotal tool in the development of negative emissions agriculture. Concurrently, there is growing interest in the use of microbial consortia inoculants to enrich composts, offering promising benefits for both soil fertility and microbial resilience under changing climatic conditions. Additionally, the valorization of animal manure into stabilized organic fertilizers reinforces nutrient cycling by effectively linking livestock and cropping systems, thereby operationalizing the foundational principles of a circular bioeconomy.

3.2.2 Climate Relevance in the South African Context

The South African agricultural landscape particularly in semi-arid provinces such as Limpopo and the Northern Cape faces persistent challenges related to declining soil fertility and limited moisture retention. Organic fertilization strategies are particularly well-suited to these conditions, as they enhance soil water-holding capacity, increase organic matter content, and improve resistance to erosion, all of which are essential attributes for climate-resilient agriculture. Despite the evident ecological benefits, this review identifies a significant gap in empirical research and field validation of such interventions within South Africa's diverse agroecological zones.

The adoption of organic amendments remains low, primarily due to the absence of standardized dosage and formulation protocols, limited availability of extension services trained in organic input application, and underdeveloped market mechanisms for the production and distribution of locally sourced organic inputs. Furthermore, there is a critical lack of field-scale trials conducted under actual South African soil and climatic conditions, and minimal application of Life Cycle Assessment (LCA) methodologies to assess the net climate impacts of organic versus synthetic fertilizers. Addressing these deficiencies requires the development of policy frameworks and economic incentives aimed at integrating organic fertilizer production into national agricultural subsidy schemes.

3.3 Thematic Cluster 2: Bioenergy and Circular Bioeconomy Valorization

The valorization of agricultural biomass for bioenergy production and bio-based materials represents one of the most strategically important intersections of climate change mitigation and rural sustainability. This thematic cluster synthesizes innovations that reframe agricultural residues, not as environmental liabilities but as renewable resources with the potential to fuel a low-carbon transition in energy and materials systems. Within the context of South Africa's energy-water-food nexus, bioenergy solutions present a particularly relevant pathway. With high biomass availability from sugarcane, maize, livestock waste, and agro-industrial residues, the country holds untapped potential for decentralized, climate-resilient energy production.

3.3.1 Bioenergy Systems: Circular Innovation and Economic Transformation

Gupta et al. (2022) and Koul et al. (2022) document significant advancements in thermochemical such as pyrolysis, gasification, and biochemical such as anaerobic digestion, fermentation conversion technologies applied to lignocellulosic biomass such as maize stover, sugarcane bagasse, and sawdust. These processes produce biogas, bioethanol, and biochar, which deliver multiple environmental and socio-economic co-benefits, including the displacement of fossil fuels in rural energy systems, the avoidance of methane emissions from unmanaged biomass decomposition, and the creation of high-value byproducts such as bioplastics and organic acids.

Karimi Alavijeh and Yaghmaei (2016), in their comparative techno-economic assessment, demonstrate that crop-specific biomass-to-biofuel systems can be feasible when logistics constraints are addressed. Their study advocates for modular biorefinery infrastructure, emphasizing small- to medium-scale facilities integrated into agricultural zones to mitigate feedstock transport limitations while generating rural employment. From a climate mitigation standpoint, integrated bioenergy systems are capable of reducing GHG emissions by 60–85% relative to fossil energy baselines (Gupta et al., 2022). The incorporation of biochar into soils further enhances the mitigation potential by transforming the system into a carbon sink an approach consistent with the negative emissions pathways outlined in the IPCC's Sixth Assessment Report (AR6).

What distinguishes this cluster is its alignment with circular bioeconomy principles, wherein agricultural waste streams are valorized into renewable energy carriers, soil amendments, and industrial feedstocks. Innovative applications include the development of agro-residue-based bioplastics as alternatives to petroleum-derived polymers (Koul et al., 2022), the deployment of decentralized bioethanol platforms to diversify off-grid energy sources for cooking and transport, and the formulation of biochar-enhanced compost products that sequester carbon while restoring soil productivity. These transitions are not only technically viable but also economically transformative, especially when integrated into cooperative-based models or public–private partnerships that address both financial and innovative gaps.

3.4 Policy and Contextual Relevance in South Africa

Despite the demonstrable benefits of bioenergy systems, South Africa has made limited progress in mainstreaming these innovations into its rural development and climate mitigation strategies. The Renewable Energy Independent Power Producer Procurement Programme (REIPPPP), while successful in mobilizing investments in wind and solar energy, has largely overlooked biomass-based energy systems. Key impediments include the capital-intensive nature of bio-processing technologies, the absence of coordinated feedstock logistics infrastructure, insufficient integration of biomass energy pathways within national climate frameworks such as the Low Emissions Development Strategy (2020), and regulatory ambiguities concerning ownership and energy use of agricultural biowaste. Given South Africa's rising energy insecurity and increased vulnerability to climate shocks, a decentralized bioenergy strategy, grounded in circular economy principles, holds promise for simultaneously enhancing energy access, reducing emissions, and promoting rural livelihoods.

3.5 Research and Implementation Gaps

This cluster also reveals substantial methodological and implementation-related gaps. Only one study to date has applied Life Cycle Assessment (LCA) methodologies to bioenergy systems within a Sub-Saharan African context. Furthermore, no studies have examined water-energy-emissions trade-offs across different crop types or regional settings. The gender and equity dimensions of bioenergy entrepreneurship, particularly regarding access to finance, land, and decision-making, remain underexplored. To address these deficits in the South African context, future research should prioritize localized techno-economic modeling of biomass value chains, integration of spatial GIS tools to identify biomass production hotspots, design of incentive structures for micro-biorefineries and climate-smart cooperative models, and the adoption of co-design approaches that actively engage farmers, energy developers, and municipal authorities in the planning and implementation processes.

3.6 Thematic Cluster 3: Vermitechnology and Biowaste Bioconversion

Vermitechnology, the biological conversion of organic waste through earthworms, has emerged as a promising low-carbon, low-input innovation at the intersection of climate adaptation, agroecological regeneration, and circular waste valorization. This cluster explores the role of vermicomposting in transforming agricultural residues into nutrient-rich organic fertilizers while contributing to GHG mitigation and soil resilience in vulnerable farming systems (Frontiers, 2022; Mupambwa et al., 2023; Robatjazi, 2020).

As a biological treatment technology, vermicomposting operates at ambient temperature, requires minimal external energy input, and is inherently decentralized (Frontiers, 2022; Robatjazi, 2020). These characteristics make it particularly suited to resource-constrained, smallholder agricultural contexts such as those prevalent in rural South Africa, where climatic variability, input costs, and soil degradation coalesce to threaten productivity and livelihoods (DEFF, 2020; Frontiers, 2022).

3.6.1 Bioconversion Efficiency and Climate Co-Benefits

Singh et al. (2020) demonstrate that vermicomposting significantly accelerates the stabilization of organic matter, reduces microbial pathogen loads, and enhances the bioavailability of key macronutrients such as nitrogen (N), phosphorus (P), and potassium (K). Their findings further indicate marked improvements in soil microbial biomass, enzymatic activity, and water retention capacity all of which are critical factors for increasing agroecosystem resilience against climate-induced stressors such as drought and soil erosion. Complementing this, Yuvaraj et al. (2021) highlight the adaptive potential of cattle dung-based vermicompost, especially under rainfed and semi-arid conditions. Their research underscores the role of feedstock selection in optimizing microbial diversity and nutrient mineralization dynamics, which influence the overall agronomic performance and ecological compatibility of vermicompost.

As Abioye et al. (2025) indicated from a climate mitigation perspective, vermitechnology offers substantial co-benefits. It reduces methane (CH_4) and nitrous oxide (N_2O) emissions commonly associated with conventional organic waste decomposition and open dumping practices. Furthermore, it diminishes the energy demand typically required for the production and transportation of synthetic fertilizers and lowers the volume of organic waste destined for landfills, thereby easing pressure on both urban and rural waste management systems (Nigussie et al., 2016a, b). Although few studies have provided precise quantifications of greenhouse gas (GHG) reductions attributable to vermitechnology, conservative estimates based on IPCC Tier 1 methodologies suggest that substituting synthetic nitrogen fertilizers with vermicompost could yield GHG reductions of approximately 0.5 to 2.5 tCO_2-equivalent per hectare per cropping season, depending on the application rate and cropping context (Abioye et al., 2025; IPCC, 2006; Nigussie et al 2016a, b).

3.6.2 Ecological and Circular Innovation Potential

Vermitechnology is uniquely positioned as an ecologically sound and circular innovation. It functions as a closed-loop nutrient recovery system, wherein organic waste streams, including livestock manure, food residues, and crop by-products, are reintegrated into the agricultural production cycle (Aslam et al., 2021; Joseph et al., 2020; Sinha et al., 2010). Earthworms play a pivotal role as bioengineers, facilitating the transformation of unprocessed biomass into stable, pathogen-free compost that enhances soil biological integrity. Vermicompost serves not only as a nutrient source but also as a plant growth promoter and biological conditioner, supporting integrated pest and disease management strategies and reducing the need for chemical inputs (Brickends Farm, 2020; PMC, 2013).

This technology is also inherently adaptable to local contexts. It operates independently of grid electricity, making it suitable for off-grid rural areas. Its design and operation are compatible with non-mechanized, low-literacy farming systems, and it promotes labour-intensive, decentralized organic waste management. These attributes open pathways for rural employment generation, particularly among women and youth, aligning vermitechnology with broader objectives of inclusive green development and community-based climate resilience (Kozenko et al., 2020; Popiha Bordolo, 2023).

3.6.3 Adoption Barriers and Knowledge Gaps

Despite its evident benefits, the adoption of vermitechnology in South Africa remains limited. A key constraint is the general lack of awareness and technical training in the areas of worm ecology, feedstock conditioning, and moisture management. This gap is compounded by the absence of well-equipped extension services capable of guiding farmers through the setup and maintenance of vermicomposting systems. Additionally, empirical data on species-specific bioconversion rates, nutrient cycling patterns, and long-term field performance under diverse South African agro-climatic zones remain scarce. Importantly, neither Singh et al. (2020) nor Yuvaraj et al. (2021) offer regionally disaggregated life cycle assessments (LCAs), which are essential for benchmarking environmental performance and informing evidence-based policy integration.

3.7 Policy and Programmatic Opportunities

Given its ecological, economic, and climate co-benefits, vermitechnology holds strong potential for inclusion in climate-smart agriculture programmes, national organic waste management strategies, green job creation plans, and agroecological extension curricula. Its integration into pilot programmes such as community-based farming initiatives, school-led agricultural training modules, and municipal organic waste valorization schemes could amplify both environmental and socio-economic outcomes (Ng'ang'a et al., 2019; Vellaiyan & Gurusamy, 2018). When coupled with complementary practices like rainwater harvesting, organic mulching, and crop diversification, vermitechnology serves as a critical component in building low-carbon, climate-resilient agricultural systems.

3.8 Thematic Cluster 4: Composting and Manure Management Systems

Composting and manure management systems (MMS) constitute foundational strategies for climate-resilient agricultural waste governance. These systems convert organic waste, particularly livestock-derived residues, into stabilized, pathogen-free amendments, thereby reducing methane emissions, mitigating eutrophication risks, and enhancing soil fertility. In the context of climate change, these strategies deliver triple dividends: they lower greenhouse gas emissions, improve the adaptive capacity of farming systems, and promote nutrient circularity within agroecosystems. This cluster synthesizes empirical evidence from diverse farming contexts to explore both low-tech, decentralized composting solutions and advanced, integrated MMS suited to large-scale operations (Ng'ang'a et al., 2019; Vellaiyan & Gurusamy, 2018).

3.8.1 *Composting as a Climate Mitigation Lever*

Pergola et al. (2018) investigated on-farm composting that integrates livestock waste and crop residues. Their findings indicate that composting under controlled aerobic conditions reduces methane emissions by over 90% relative to traditional open-pit manure storage. The study employed Life Cycle Assessment (LCA) to demonstrate the comparative benefits of composting in terms of energy savings, reduced synthetic fertilizer demand, and avoided emissions from raw manure volatilization. Further, compost application contributes to long-term soil carbon storage, indirectly supporting climate mitigation through enhanced sequestration. Well-matured compost improves soil structure, water retention, and microbial biodiversity. All critical attributes for withstanding prolonged dry periods, extreme rainfall, and nutrient leaching events associated with climate instability Varma et al., 2021).

3.8.2 *Advanced Manure Management Systems and Biogas Integration*

Varma et al. (2021) broaden the scope of composting research by analyzing integrated Manure Management Systems (MMS) in large-scale dairy and swine production contexts. Their study incorporates advanced processes such as anaerobic digestion, thermal drying, and mechanical separation, which collectively yield stabilized organic amendments alongside biogas for renewable energy generation. The climate relevance of these systems is considerable. Anaerobic digestion of livestock manure has the potential to abate up to 2.5 tonnes of CO_2-equivalent per tonne of volatile solids, contingent on system design and operational efficiency (IPCC, 2019). In addition to emissions reduction, the captured biogas serves as a clean energy substitute, displacing the use of firewood, liquefied petroleum gas (LPG), or coal in household and industrial applications. Furthermore, the processed effluent from digestion systems can be recycled for irrigation, promoting water-efficient agriculture in regions facing increasing climatic water stress. Nevertheless, the widespread implementation of such technologies in rural South Africa is hindered by significant capital requirements, the need for sophisticated technical expertie, and limited institutional support structures (Brickends Farm, 2020).

3.8.3 Innovation and Circularity in Composting Systems

Recent innovations within the composting cluster highlight the potential for increased circularity and nutrient recovery. Notable developments include the application of compost accelerants, such as thermophilic microbes and microbial inoculants, which significantly reduce composting time while enhancing nutrient density. Other advancements involve blended composting systems that integrate livestock manure with agro-industrial residues such as sugarcane pressmud and fruit processing waste thereby enhancing organic content and diversifying feedstock inputs. Additionally, the incorporation of biochar into MMS has proven effective in reducing nutrient volatilization losses and improving compost stability and carbon retention. These integrated systems contribute meaningfully to circular bioeconomy frameworks by reinserting organic waste into productive use cycles, minimizing dependency on external inputs, and closing nutrient loops essential for sustainable farming systems.

3.8.4 Contextual Relevance for South Africa

In the South African context, provinces such as the Free State, Eastern Cape, and KwaZulu-Natal host intensive livestock operations that generate substantial quantities of manure annually. However, much of this organic waste is disposed of through informal and unregulated means, leading to uncontrolled methane emissions, nitrate leaching, and water contamination (Moeletsi & Tongwane, 2015; Moleko, 2021). The reviewed literature identifies context-specific adoption pathways. For example, smallholder farmers may implement low-cost aerobic composting techniques using basic windrow systems and manual turning practices, which are both accessible and resource-efficient. On the other hand, commercial livestock operations are better positioned to transition towards anaerobic digestion systems, particularly when supported by life cycle assessment (LCA)-based decision tools and market instruments such as carbon credit mechanisms ((Dell'Orto and Trois, 2024). Despite these opportunities, South Africa currently lacks a comprehensive regulatory or incentive framework that formally recognizes compost production as a climate mitigation service. Compost remains undervalued within the national soil amendment market, even though its environmental and agronomic benefits are well-documented ((Dell'Orto and Trois, 2024).

3.8.5 *Barriers and Strategic Entry Points*

Several structural barriers continue to impede the scaling of composting and biogas systems in South Africa. Chief among these is the absence of financial incentives, such as compost production subsidies or feed-in tariffs for biogas energy. Infrastructure for composting remains scarce in rural areas, and technical capacity is insufficient for ensuring compost quality control and for monitoring emissions during decomposition processes. Despite these challenges, strategic entry points for scaling exist. Integrating composting practices into municipal organic waste diversion schemes presents an immediate opportunity for expanding impact. Establishing community-based composting hubs particularly those linked to school gardens and smallholder farmer cooperatives can generate multiple co-benefits, including education, food security, and livelihood support. Moreover, leveraging international carbon finance mechanisms could catalyze transitions from traditional manure disposal to biogas generation systems, aligning rural waste management with climate-smart development pathways.

3.8.6 *Comparative Synthesis of Thematic Clusters*

To contextualize the relative performance of composting and MMS among the broader set of climate-compatible AWM strategies, Fig. 3 presents a comparative evaluation across four key dimensions: GHG mitigation potential, soil health benefits, cost-efficiency (inverted), and scalability. This synthesis supports evidence-based prioritization for policy and implementation in South African agriculture.

3.9 *Environmental and Socio-Economic Implications*

The thematic clusters reviewed in this study demonstrate that sustainable agricultural waste management (AWM) strategies deliver co-benefits across environmental, economic, and social dimensions. These synergies position AWM not merely as a waste minimization measure but as a multifunctional climate solution, particularly well-suited to countries like South Africa that face concurrent challenges of ecological degradation, food insecurity, and rural poverty.

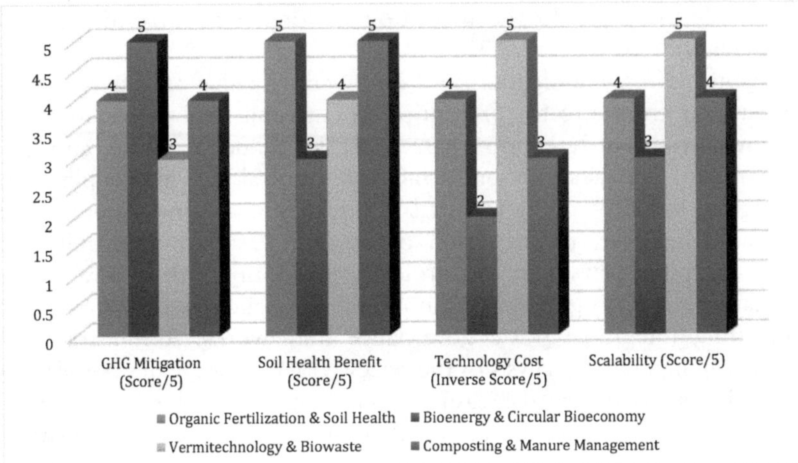

Fig. 3 Comparative evaluation of climate-compatible agricultural waste management strategiesComparative performance of four agricultural waste management strategies across key climate-relevant criteria. Scores (1–5) reflect synthesized evidence on GHG mitigation, soil health, cost-efficiency (inverted), and scalability, based on representative studies from thematic clusters.

3.9.1 Environmental Benefits: GHG Mitigation, Soil Regeneration, and Pollution Abatement

A key environmental advantage of climate-compatible agricultural waste management (AWM) strategies lies in their capacity to mitigate greenhouse gas (GHG) emissions through multiple pathways. Methane (CH_4) emissions, commonly associated with anaerobic manure storage, landfilling, and open biomass burning, are significantly reduced through the adoption of composting and anaerobic digestion technologies (Pergola et al., 2018; Varma et al., 2021). Similarly, nitrous oxide (N_2O) emissions often resulting from the overuse or improper application of synthetic fertilizers can be curtailed by integrating organic amendments such as biochar and vermicompost into nutrient management regimes (Bhunia et al., 2021; Zhang et al., 2021). In addition to emissions abatement, these interventions promote carbon sequestration in soils by incorporating stabilized organic matter, particularly when biochar is co-applied, thereby contributing to the formation of long-term terrestrial carbon sinks (Harindintwali et al., 2021).

Beyond climate mitigation, AWM practices enhance soil regeneration and contribute to the prevention of environmental degradation. By improving nutrient retention, these strategies reduce leaching, eutrophication, and soil acidification, which are critical for maintaining the ecological integrity of agricultural landscapes and downstream water systems. Moreover, the diversion of organic waste from landfills and open dumping plays a pivotal role in promoting urban–rural waste circularity, alleviating pressure on municipal solid waste systems and supporting integrated waste governance frameworks. As illustrated in Fig. 4, a comparative analysis of the reviewed AWM strategies reveals that while all approaches yield measurable environmental benefits, they vary in their relative effectiveness across dimensions such as GHG mitigation, soil health improvement, and cost accessibility. These variations underscore the importance of context-specific implementation strategies that are tailored to maximize co-benefits under varying climatic, agronomic, and socio-economic conditions.

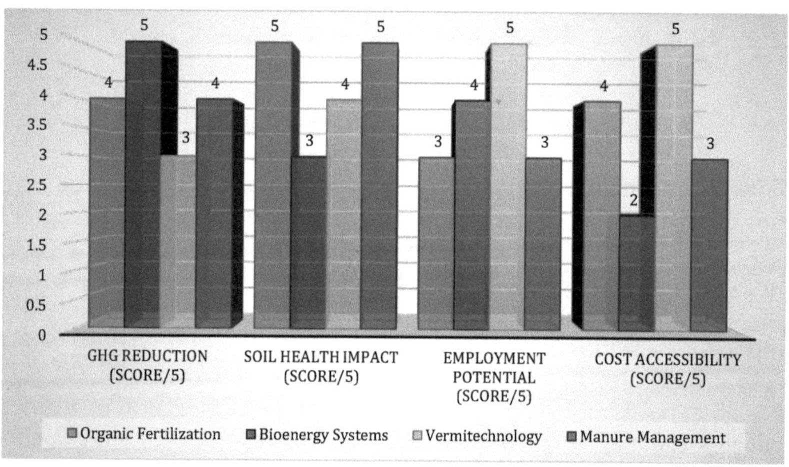

Fig. 4 Comparative co-benefits of AWM strategies across GHG mitigation, soil health, employment, and cost accessibility. Scores (1–5) are synthesized from representative studies

3.9.2 *Socio-Economic Benefits: Resilient Livelihoods and Rural Value Chains*

The reviewed agricultural waste management (AWM) strategies offer substantial socio-economic benefits, particularly by reducing input costs for farmers through the substitution of synthetic fertilizers with locally produced organic alternatives. These strategies not only contribute to cost savings but also improve crop yields and resilience by enhancing nutrient uptake, root development, and microbial symbiosis in soils. Employment creation is another significant co-benefit, particularly in sectors such as vermicomposting, biogas plant maintenance, and compost marketing—areas well-suited to microenterprise development in rural contexts. Studies by Gupta et al. (2022) and Yuvaraj et al. (2021) underscore the entrepreneurial potential of decentralized waste valorization systems. Small-scale biogas plants, vermicompost units, and biofertilizer cooperatives present scalable models for inclusive green growth, especially when embedded within public–private partnerships or supported by climate finance instruments such as green bonds and results-based financing. In the South African context, where youth unemployment and smallholder exclusion from value chains remain pervasive, AWM can be strategically linked to skills development programmes, particularly those focusing on composting and bioconversion rural innovation hubs, and gender-equity interventions. This is especially pertinent given that many organic input enterprises are led by women in comparable contexts across the Global South.

3.9.3 *Equity, Accessibility, and the Just Transition Imperative*

A central insight from the synthesis is that while advanced technologies such as anaerobic digestion and biorefinery systems provide significant emissions reduction potential, they are typically capital-intensive and inaccessible to resource-constrained smallholders. In contrast, low-input strategies such as vermitechnology and windrow composting are more accessible and deliver immediate benefits, positioning them as essential components of a just transition in climate-smart agricultural policy. To ensure equitable access and scalable adoption, national and municipal stakeholders must develop tiered support mechanisms tailored to the needs of small-, medium-, and large-scale producers. Incorporating waste-to-resource solutions into climate-smart agricultural subsidies, while simultaneously enabling access to low-interest green finance and carbon revenue streams, can significantly enhance the viability and appeal of community-based AWM initiatives (Dell'Orto & Trois, 2024).

3.9.4 *Cross-Cutting Challenges and Caveats*

Despite their many advantages, AWM strategies face several implementation challenges. Empirical studies often fail to provide comprehensive life cycle emissions data and cost–benefit analyses, thereby limiting their utility in national carbon budgeting processes. Additionally, cultural perceptions and a lack of awareness impede farmer adoption, particularly in contexts where waste reuse is either stigmatized or poorly understood. The absence of enabling policies, appropriate market incentives, and integrated decision-support tools further constrains the institutionalization of AWM at scale.

These limitations highlight the critical need for systemic policy integration, localized adaptation of innovations, and multi-stakeholder coordination to embed AWM meaningfully within broader climate resilience planning frameworks. The literature reveals a marked shift in agricultural waste management paradigms from isolated disposal practices towards integrated, innovation-driven systems that align with climate mitigation and adaptation goals. This transition is characterized by the convergence of technological innovation, ecological design principles, and circular bioeconomy frameworks, which collectively redefine how agricultural residues are managed and valorized.

3.9.5 *Bio-Based Innovation in Waste Valorization*

One prominent trend is the expansion of agricultural waste valorization into bio-industrial domains. Organic residues are increasingly being processed into high-value commodities such as biochar, bioplastics, bioethanol, and organic acids using thermochemical and biochemical conversion platforms (Gupta et al., 2022; Koul et al., 2022). These processes enable the co-production of renewable energy and soil amendments, thereby enhancing resource efficiency and supporting regenerative land use. Such innovations are foundational to the circular bioeconomy, where waste is transformed into renewable inputs, reducing dependency on petrochemical-derived products and imported fertilizers, particularly important in resource-constrained economies like South Africa.

3.9.6 *Integration of Biological Enhancements and Soil Biotech*

Another significant development involves the bio-enrichment of compost and organic fertilizers with beneficial microbial consortia. The use of nitrogen-fixing bacteria, phosphate-solubilizing microbes, and thermophilic composting accelerants has been shown to improve nutrient bioavailability and reduce gaseous losses during composting processes (Harindintwali et al., 2021). These biological enhancements contribute to the restoration of soil microbiomes, which is essential for long-term carbon sequestration and for supporting adaptive crop responses under increasingly variable climatic conditions.

3.9.7 *Digitalization and Precision Composting*

Digital technologies are beginning to shape the future of high-efficiency composting systems. Precision composting platforms now incorporate real-time monitoring of variables such as temperature, pH, and moisture content, alongside automated turning systems and microbial diagnostics. The application of Internet of Things (IoT) devices and remote sensing technologies is enabling spatial tracking of organic waste flows and optimizing compost production processes. Although currently limited to industrial-scale or high-tech farming operations, these digital tools hold potential to improve the accuracy of life cycle assessments, streamline carbon accounting, and enhance the scalability of climate-smart composting practices.

3.9.8 *Semi-Industrial Vermitechnology Platforms*

Vermitechnology is undergoing a transformation from small-scale backyard systems to semi-industrial platforms. These advanced systems incorporate pre-treatment stages such as shredding and moisture adjustment, utilize optimized earthworm species for specific feedstocks, and deploy modular vermireactors capable of operating in batch or continuous flow modes (Yuvaraj et al., 2021). Such innovations are tailored to meet rising demand for high-quality organic fertilizers, particularly in peri-urban and commercial horticultural markets. The emergence of branded vermicompost products, with verified nutrient compositions and microbial activity indices, may soon become part of the organic certification framework for climate-smart agriculture.

3.9.9 Tool-Based Decision Support and Life Cycle Integration

There is a growing emphasis on the use of Decision Support Tools (DSTs) and Life Cycle Assessment (LCA) to guide the selection of appropriate technologies based on local environmental and economic conditions. Pergola et al. (2018) and Varma et al. (2021) emphasize the importance of comparing emissions profiles, energy use, and nutrient balance across different systems, as well as integrating economic and ecological indicators to support investment decisions under uncertainty. However, the application of regionally calibrated LCA models remains limited in Sub-Saharan Africa, underscoring the need for accessible, open-source decision tools to facilitate evidence-based planning and stakeholder engagement.

3.9.10 Convergence of Climate Policy and Waste Management

There is increasing recognition of the strategic role that AWM can play in achieving national climate targets and sustainable development objectives. Innovations such as composting, bioenergy, and vermitechnology directly contribute to greenhouse gas mitigation in agriculture (IPCC sectors 3A and 3B), improve soil carbon stocks under climate-smart land management practices, and align with the sustainable biomass use principles set forth in the Global Bioenergy Partnership (GBEP). Nevertheless, integration into climate finance mechanisms and regulatory frameworks remains fragmented in the South African context. There is a pressing need to codify AWM as a climate mitigation category within national greenhouse gas inventories, establish carbon crediting mechanisms for verified composting and bioenergy projects, and incorporate AWM metrics into the implementation of the Climate Change Bill and the Just Transition Framework. These developments signify a broader systems transformation in AWM, where technological advancement must be matched by governance innovation. As illustrated in Fig. 5, achieving climate-resilient agricultural outcomes requires coordinated alignment of enabling policies, financing instruments, farmer participation, and institutional capacity to convert inputs into long-term environmental and livelihood impacts.

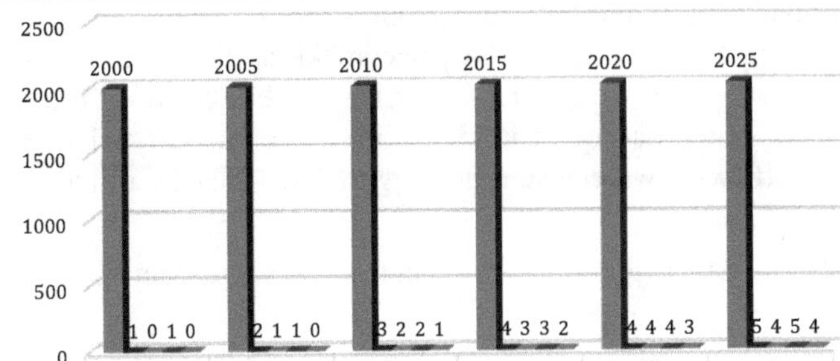

	1	2	3	4	5	6
Year	2000	2005	2010	2015	2020	2025
Composting	1	2	3	4	4	5
Vermitechnology	0	1	2	3	4	4
Bioenergy	1	1	2	3	4	5
Biochar Fertilization	0	0	1	2	3	4

Year Composting Vermitechnology Bioenergy Biochar Fertilization

Fig. 5 Theory of change for climate-smart agricultural waste management

4 Discussion

4.1 Thematic Gaps in Climate-Compatible Agricultural Waste Management Literature

While the literature review demonstrates increasing innovation and momentum towards climate-aligned agricultural waste management (AWM), it also reveals significant methodological, empirical, and systemic gaps that constrain progress, especially within the context of South African agroecosystems.

1. Underrepresentation of Empirical and Region-Specific Data

Despite the ecological and technological diversity of the reviewed studies, few offered long-term field trials, climate impact modeling, or adaptive performance assessments tailored to South Africa's heterogeneous agro-climatic zones. Composting and vermicomposting studies (e.g., Bhunia et al., 2021; Singh et al., 2020) often rely on generalized soil or climate conditions, overlooking local variables such as soil mineralogy, microbial ecology, and seasonal precipitation volatility. This lack of regionally calibrated data undermines the transferability of global findings and hampers evidence-based policymaking. For instance, while biochar-enhanced composting shows emission reduction potential, no South African study has quantified its life cycle carbon balance or modelled trade-offs with water-use efficiency or crop-specific outcomes.

2. Limited Integration of Climate Tools and Metrics

Only a minority of studies employed Life Cycle Assessment (LCA) or Decision Support Tools (DSTs) capable of guiding context-specific AWM interventions under uncertainty. Without these tools, interventions may fail to account for unintended consequences, such as excess phosphorus runoff, nutrient imbalances, or energy intensity of pre-treatment stages. Moreover, GHG quantification frameworks are inconsistently applied. Where present, most rely on IPCC Tier 1 default coefficients rather than direct emissions measurements, soil flux monitoring, or system-wide modeling approaches needed for robust mitigation benchmarking.

3. Fragmented Socio-Institutional Integration

Technological innovation in agricultural waste management (AWM) has progressed more rapidly than institutional adaptation. The review identifies a notable lack of attention to critical socio-institutional dimensions that influence the uptake and scaling of AWM technologies. Specifically, there is limited engagement with farmer knowledge systems and the behavioural incentives that drive decision-making around waste reuse. Cultural perceptions, particularly those surrounding the use of manure and vermicompost, are often overlooked, despite their significant influence on adoption patterns. Furthermore, the roles of extension services, agricultural cooperatives, and municipal actors in facilitating technology dissemination and support remain underexplored in the literature. For example, although vermicomposting is widely regarded as a socially inclusive and scalable practice, its uptake continues to be constrained by institutional fragmentation, weak market linkages, and a lack of access to microfinance facilities or certification mechanisms that would validate and support the use of vermicompost in formal agricultural markets.

4. Policy and Infrastructure Gaps in South Africa

Compared to frontrunners such as India, Brazil, and Germany, South Africa lacks a comprehensive policy framework that integrates agricultural waste management (AWM) into its broader climate mitigation and adaptation architecture. Although the Climate Change Bill (2022) and the Low Emissions Development Strategy (LEDS) identify agriculture as a priority sector, they fall short of explicitly addressing waste valorization or the development of bio-based circular systems. Several critical policy and institutional gaps persist. There are no targeted subsidies or carbon-based incentives to promote the production and application of compost, biogas, or vermifertilizer. Additionally, the country lacks a national organic waste registry, does not include bio-based soil amendments within its greenhouse gas inventory categories, and has not established quality standards for such amendments. Furthermore, interdepartmental coordination remains weak among the Department of Agriculture, Forestry and Fisheries (DAFF), the Department of Forestry, Fisheries and the Environment (DFFE), and municipal waste management authorities. These institutional deficiencies significantly hinder the transition from linear, high-emission waste systems to integrated, climate-compatible resource flows necessary for a low-carbon agricultural economy.

4.2 *Future Research Directions*

Despite growing recognition of agricultural waste management (AWM) as a cornerstone of sustainable agriculture and climate resilience, critical research gaps persist in socio-economic integration, technological adaptation, and environmental modeling. Future research as per Table 3 must go beyond technical efficiency to incorporate equity, inclusivity, and localized system optimization. The following thematic priorities are proposed to guide future empirical and applied investigations in South Africa and similar agroecological contexts.

Table 3 Priority areas for future research in climate-resilient agricultural waste management (AWM)

Research focus	*Key themes and proposed actions*
Cross-cutting issues	– Assess gendered impacts of AWM innovations – Evaluate employment multipliers in decentralized systems – Investigate redistributive climate financing for youth, women, and marginalized producers
Empirical optimization of organic inputs	– Design longitudinal field trials for crop- and soil-specific organic fertilizer protocols – Quantify soil carbon, microbial diversity, and moisture retention changes – Validate biochar- and microbe-enriched composts under abiotic stress such as drought, salinity
Climate impact modeling and LCA integration	– Integrate Life Cycle Assessment (LCA), Life Cycle Costing (LCC), and water-energy-emissions trade-offs – Apply GIS-based spatial modeling for optimal deployment zones – Develop open-source Decision Support Tools (DSTs) for stakeholders
Innovation Co-Design with Smallholders	– Conduct participatory research aligned with local social norms, labour, and access to resources – Explore integration of indigenous knowledge in composting and soil stewardship – Co-develop business models for low-tech solutions such as vermitechnology

5 Conclusion

This systematic literature review has critically examined the current state of climate-compatible agricultural waste management (AWM) strategies, emphasizing their applicability within the South African context. Drawing upon ten peer-reviewed studies encompassing diverse agroecological and technological domains, the review synthesized four thematic clusters organic fertilization and soil health enhancement, bioenergy and circular bioeconomy valorization, vermitechnology, and composting and manure management systems. Each cluster presents distinct climate, ecological, and socio-economic co-benefits. The findings unequivocally support the transformative potential of AWM in advancing multiple sustainability objectives. These include the mitigation of greenhouse gas emissions through reductions in methane and nitrous oxide, climate adaptation via enhanced soil water retention, fertility and microbial resilience, the establishment of circular resource flows that reduce dependence on fossil-based inputs, and inclusive economic development facilitated by labour-intensive, low-cost waste valorization systems.

However, despite these promising outcomes, the review also identifies persistent limitations that constrain the full realization of AWM's climate potential. These include a lack of empirical field validation under South African agro-climatic conditions, weak integration of life cycle assessment frameworks, decision-support tools, and climate finance mechanisms, limited policy infrastructure to support scalable and equitable adoption, and fragmented institutional coordination that hinders effective inclusion in national mitigation plans. Addressing these deficits will require a coordinated, interdisciplinary approach that bridges technological innovation, policy alignment, and inclusive governance mechanisms. Future research and implementation efforts should focus on co-producing context-specific knowledge with farmers, cooperatives, and extension services, while simultaneously mainstreaming AWM into South Africa's Nationally Determined Contributions (NDCs), the Climate Change Bill, and the Just Transition framework.

Moreover, creating enabling environments for public–private partnerships, expanding access to microfinance and carbon markets, and fostering innovation in bio-based product development, vermicomposting hubs, and decentralized biogas systems will be critical. As South Africa continues to confront escalating climate pressures and resource constraints, AWM should not be viewed merely as a waste management intervention but rather as a strategic lever for agricultural transformation, environmental restoration, and rural revitalization. When embedded within a systems-oriented framework that connects soil health, energy resilience, livelihood generation, and emissions reduction, climate-compatible AWM can serve as a foundational pillar of a low-carbon, circular, and just agricultural economy. This review contributes to that agenda by offering a scientifically rigorous and policy-relevant synthesis to inform scholars, practitioners, and policymakers in designing integrative, scalable, and context-sensitive waste management solutions for climate-resilient agriculture.

References

Abioye, O. M., et al. (2025). An overview of vermicompost's role in reducing greenhouse gas emissions, improving soil health, and crop yields. *Applied Science and Engineering Progress, 18*(2), 7586. [online] https://ojs.kmutnb.ac.th/index.php/ijst/article/download/7586/pdf_551. Accessed 27 June 2025.

Aslam, M., et al. (2021). Vermicomposting by bio-recycling of animal and plant waste: A review. *International Journal of Environment, 10*(3).

Bhunia, S., et al. (2021). Application of recycled slaughterhouse wastes as an organic fertilizer for successive cultivations of bell pepper and amaranth. *Science of the Total Environment*.

Brenya, R., Akomea-Frimpong, I., Ofosu, D., & Adeabah, D. (2022a). Barriers to sustainable agribusiness: A systematic review and conceptual framework. *Journal of Cleaner Production, 330*, Article 129812.

Brenya, R., Akomea-Frimpong, I., Ofosu, D., & Adeabah, D. (2022b). Barriers to sustainable agribusiness: A systematic review and conceptual framework. *Journal of Agribusiness in Developing and Emerging Economies*, ahead-of-print(ahead-of-print).

Brickends Farm. (2020). *Earthworms Vermicompost: A powerful crop nutrient and plant protector*. [online] https://brickendsfarm.com/wp-content/uploads/2020/03/vermicompost-soil-conditioner.pdf. Accessed 20 June 2025.

Dada, E. O. (2024). Vermiculture and vermicomposting: twin earthworm technologies, suitable for households and smallholder farmers to convert wastes to wealth—A narrative review. *FUDMA Journal of Agriculture and Agricultural Technology, 10*(1), 134–143. [online] https://doi.org/10.33003/jaat.2024.1001.17 [Accessed 15 Jun. 2025].

DEFF. (2020). *Actionable guidelines for the implementation of climate-smart agriculture in South Africa. Volume 2: Climate Smart Agriculture Practices*. Department of Environment, Forestry & Fisheries, Republic of South Africa. [online] https://www.dffe.gov.za/sites/default/files/docs/csa_volume2.pdf [Accessed 10 Jun. 2025].

Dell'Orto, A., & Trois, C. (2024). Double-stage anaerobic digestion for biohydrogen production: A strategy for organic waste diversion and emission reduction in a South African municipality. *Sustainability, 16*(16), 7200. [online] https://www.mdpi.com/2071-1050/16/16/7200. Accessed 20 June 2025.

Food and Agriculture Organization (FAO). (2022). *The future of food and agriculture: Trends and challenges*. FAO. [online] https://www.fao.org/3/i6583e/i6583e.pdf. Accessed 27 June 2025.

Frontiers. (2022). Vermicomposting as a climate-smart technology for vegetable crops. *Frontiers in Sustainable Food Systems*. [online] https://www.frontiersin.org/articles/10.3389/fsufs.2022.757792/epub. Accessed 20 June 2025.

Gupta, R., et al. (2022). Lignocellulosic biomass valorization for bioethanol production: A circular bioeconomy approach. *BioEnergy Research, 15*, 1820–1841. [online] https://link.springer.com/article/10.1007/s12155-022-10401-9. Accessed 15 June 2025.

Harindintwali, J., et al. (2021). *Study on the application of biochar to soil pollution remediation and its effect on soil microbial activity*. (excerpt) [online] https://papers.ssrn.com/sol3/Delivery.cfm/.../abstractid=4681400.papers.ssrn.com. Accessed 2 June 2025.

IEA. (2024). *World energy investment special report: Reducing the cost of capital*. International Energy Agency. [online] https://www.iea.org/reports/reducing-the-cost-of-capital. Accessed 23 June 2025.

IPCC Tier-1 methodology assigns default factors for N_2O emissions from nitrogen inputs—both synthetic and organic—enabling rough GHG reduction estimates through fertilizer substitution (IPCC, 2006)

Joseph, R., et al. (2020). Nutrient recovery and vermicompost production from livestock solid wastes with epigeic earthworms. *Bioresource Technology, 313*, 123690. [online] https://www.researchgate.net/publication/342227595_Nutrient_recovery_and_vermicompost_production_from_livestock_solid_wastes_with_epigeic_earthworms. Accessed 25 June 2025.

Kamilaris, A., Funes Mesa, I., Savé, R., De Herralde, F., & Prenafeta-Boldú, F. X. (2020). *Can animal manure be used to increase soil organic carbon stocks in the Mediterranean as a mitigation climate change strategy*? arXiv, 2007.10823. [online] https://arxiv.org/abs/2007.10823. Accessed 21 June 2025.

Karimi Alavijeh, M., & Yaghmaei, S. (2016). Biochemical production of bioenergy from agricultural crops and residue in Iran. *Waste Management, 52*, 375–394. [online] https://doi.org/10.1016/j.wasman.2016.03.025. Accessed 27 June 2025.

Khan, S. A., & Mahajan, I. (2025). *Barriers and drivers to circular economy in agribusiness: A systematic review*. Circular Economy and Sustainability. [online] Springer. https://doi.org/10.1007/s43615-024-00391-5. Accessed 25 June 2025.

Koul, S., et al. (2022). Utilization of agricultural waste biomass and recycling toward circular bioeconomy. *Environmental Science and Pollution Research*, (2022). [online] https://doi.org/10.1007/s11356-022-20669-1. Accessed 22 June 2025.

Kozenko, K., et al. (2020). *Vermicompost production as a means of inclusive green economy development in rural settlements*. ATG Proceedings. [online] https://www.researchgate.net/publication/339360725_Vermicompost_Production_as_a_Means_of_Inclusive_Green_Economy_Development_in_Rural_Settlements. Accessed 23 June 2025.

Miladinov, G. (2023). Impacts of population growth and economic development on food security in low- and middle-income countries. *Frontiers in Human Dynamics, 5*, 1–12. [online] https://www.frontiersin.org/journals/human-dynamics/articles/10.3389/fhumd.2023.1121662/full. Accessed 23 June 2025.

Moeletsi, M. E., & Tongwane, M. I. (2015). Methane and nitrous oxide emissions from manure management in South Africa. *Animals, 5*(2), 193–205. [online] https://doi.org/10.3390/ani5020193. Accessed 22 June 2025.

Moleko, K. A. (2021). *Evaluation of municipal waste services in Matjhabeng Local Municipality, Free State*. Central University of Technology thesis. [online] https://ir.cut.ac.za/bitstream/handle/11462/2234/Moleko%2C%20Kelebohile%20Andronica.pdf. Accessed 27 June 2025.

Mupambwa, H. A., Nciizah, A. D., & Gabriel, N. N. (2023). Editorial: Mixed farming technologies for increased resilience, mitigation and adaptation to a changing climate for smallholder farmers in Southern Africa. *Frontiers in Sustainable Food Systems, 6*, 1082137. [online] https://doi.org/10.3389/fsufs.2022.1082137. Accessed 26 June 2025.

Ng'ang'a, J. K., et al. (2019). Vermicomposting as a climate-smart agricultural strategy in Kenya: The nexus between climate change, research and innovation. *ResearchGate*. [online] https://www.researchgate.net/publication/341900465_Vermicomposting_for_climate_smart_agriculture. Accessed 24 June 2025.

Nigussie, et al. (2016a). *Report vermicomposting can reduce N_2O emissions by 23–36% and CH_4 by 22–26% compared to conventional composting, translating into notable CO_2-equivalent savings under field conditions*.

Nigussie, D., et al. (2016b). Vermicomposting as a technology for reducing nitrogen losses and greenhouse gas emissions from small-scale composting. *ResearchGate*. [online] https://www.researchgate.net/publication/306128308_Vermicomposting_as_a_technology_for_reducing_nitrogen_losses_and_greenhouse_gas_emissions_from_small-scale_composting. Accessed 27 June 2025.

Pergola, M., et al. (2018). Composting as manure disposal strategy in small/medium-size livestock farms: some demonstrations with operative indications. *Sustainability, 12*(8), 3315. [online] https://www.mdpi.com/2071-1050/12/8/3315. Accessed 15 June 2025.

Popiha Bordoloi, P. (2023). Vermicompost: A tool for agripreneurship development for the unemployed youth of India. *Research Trend*. [online] https://www.researchgate.net/publication/376679254_Vermicompost_A_Tool_for_Agripreneurship_Development_for_the_Unemployed_Youths_of_India. Accessed 10 June 2025.

Robatjazi, J. (2020). An overview on vermicompost environmental impacts (from past to future). *Agri & Aquac. Biochemistry & Systems Biology*. [online] https://www.longdom.org/open-access/an-overview-on-vermicompost-environmental-impacts-from-past-to-future-103775.html. Accessed 12 June 2025.

Robatjazi, J. (2023). An overview on vermicompost environmental impacts (from past to future). *Agri & Aquac. Biochemistry & Systems Biology, 14*(4). [online] https://www.longdom.org/open-access/an-overview-on-vermicompost-environmental-impacts-from-past-to-future-103775.html. Accessed 11 June 2025.

Ryan, S. J., Pusey, A. E., Gitau, M., Dubach, J. M., Turner, J. W., & Brashares, J. S. (2014). Population pressure and global markets drive forest change in Africa's Albertine Rift. *Environmental Research Letters, 9*(4), 044006. [online] https://iopscience.iop.org/article/10.1088/1748-9326/9/4/044006. Accessed 7 June 2025.

Singh, K., & Sinha, R. (2020). *Vermicomposting: A gateway to sustainable agriculture production*. Conference paper.

Sinha, R. K., et al. (2010). Vermiculture technology: Reviving the dreams of Sir Charles Darwin.... *Technology and Investment, 1*(3), 155–172. [online] https://www.scirp.org/journal/paperinformation?paperid=2490. Accessed 3 June 2025.

Smith, P., Taghizadeh-Toosi, A., Rieu, M., et al. (2023). Biochar-co-compost: A promising soil amendment to restrain GHG emissions during composting. *Agronomy, 14*(7), 1583. [online] https://www.mdpi.com/2073-4395/14/7/1583. Accessed 5 June 2025.

Taghizadeh-Toosi, A., Clough, T. J., Condron, L. M., O'Callaghan, M., Sherlock, R. R., & Schmidt, M. (2021). Biochar addition reduces non-CO_2 greenhouse gas emissions in cattle-manure compost. *Journal of Environmental Quality, 50*(1), 1–10. [online] https://pubmed.ncbi.nlm.nih.gov/37117005/. Accessed 6 June 2025.

Teklu, Y., Adekunle, A. A., & Becker, M. (2021). The current status, challenges and prospects of using biomass energy in Ethiopia. *Biotechnology for Biofuels, 14*: 123. [online] https://link.springer.com/article/10.1186/s13068-021-02060-3. Accessed 27 June 2025.

Varma, V., et al. (2021). Anaerobic digestion of animal manure and influence of organic and operational factors on methane yield and greenhouse gas emissions. *Frontiers in Energy Research*. https://www.frontiersin.org/journals/energy-research/articles/10.3389/fenrg.2021.740314/full. Accessed 21 June 2025.

Vellaiyan, D., & Gurusamy, P. (2018). Impact of vermicomposting training programme on production, economics and employment generation of farmer—A case study. *Asian Journal of Agricultural Extension, Economics & Sociology, 27*(2), 1–5. https://www.frontiersin.org/articles/10.3389/fsufs.2022.757792/epub. Accessed 21 June 2025.

Yuvaraj, P., et al. (2021). *Centrality of cattle solid wastes in vermicomposting technology—A cleaner resource recovery and biowaste recycling option for agricultural and environmental sustainability.*

Zhang, Y., et al. (2021). *Biochar has large specific surface area that helps improve soil properties and soil enzyme activity* (citing Zhang et al. 2021). [online] https://onlinelibrary.wiley.com/doi/10.1111/gcbb.12995. Accessed 21 June 2025.

CHAPTER 2

Harnessing Indigenous Leafy Vegetables for Poverty Alleviation and Economic Upliftment in Marginalized South African Communities

Manaka J. Makgato, Salmina Mokgehle, Azwimbavhi Reckson Mulidzi, Willem Jansen van Rensburg, Mariette Truter, Michael Bairu, Fhatuwani Nixwell Mudau, and Hintsa Araya

1 Introduction

The prevalence of climate change increase and variability globally has opened a window of opportunity in research on the use of indigenous food crops to alleviate hunger and poverty, and this is evident in Sub-Saharan regions of Africa (Bokelmann et al., 2022; Chanza &

M. J. Makgato · A. R. Mulidzi
Soil and Water Science Division, Agricultural Research Council, Infruitec-Nietvoorbij, Cape Town, South Africa
e-mail: MakgatoM@arc.agric.za

N. Tshikovhi et al. (eds.), *The BRICS Imperative*,
https://doi.org/10.1007/978-3-032-05932-1_2

Musakwa, 2022; Modi & Mabhaudhi, 2016). According to the Department of Trade, Industry and Cooperation of the Republic of South Africa (DTIC, 2023), food demands in Africa continue to outstrip local supply, with the African continent spending approximately thirty-five billion in US$ annually on food imports to supplement local deficits. Commercial crop advances (agronomy and breeding) are under threat from climate change, land degradation and erratic rainfalls, and unlike other commercial crops, which are adapted to different regions, indigenous food crops are uniquely suited to local environments and are more resilient to climatic variations and tolerant to local pests and diseases (Bokelmann et al., 2022; Mabhaudhi et al., 2017; Nhamo et al., 2022).

The agricultural expertize of indigenous communities is often overlooked by policy and decision-makers (Mudau et al., 2022), and interventions based on mainstream crops and external technologies may fail indigenous communities where a vast range of crops are cultivated in diverse production systems and marginal environments (Qwabe & Pittawaty, 2023). Hunger, malnutrition, and poverty in indigenous communities are high. The challenges should be approached from the perspectives and resources of indigenous people. Marginal environments and the indigenous communities who cultivate these indigenous food crops share a common reality—they are often overlooked and forgotten. Their soils and climates, crops and livestock, beliefs and knowledge systems rarely attract academic interest, policy studies or investment (Azam-Ali et al., 2023).

M. J. Makgato
Department of Agriculture and Animal Health, University of South Africa (UNISA), Pretoria, South Africa

W. J. van Rensburg · M. Truter · M. Bairu · H. Araya
Agricultural Research Council-Vegetables, Industrial and Medicinal Plants (ARC- VIMP), Pretoria, South Africa

S. Mokgehle (✉)
School of Agricultural Sciences, University of Mpumalanga, Mbombela, South Africa

F. N. Mudau
School of Agricultural, Earth and Environmental Sciences, University of KwaZulu-Natal, Durban, South Africa

Several reasons are advanced for neglecting these indigenous leafy vegetables despite their higher nutrient contents compared to the conventional (Gerrano, 2021). Indigenous leafy vegetables are known to be well-adapted to the climatic and natural resource conditions of the native environment (Alulu et al., 2023; Mabhaudhi et al., 2019; Omotayo & Aremu, 2020). These crops hold a significant place in the socio-cultural lives of communities, and enhancing research and innovation focused on them is considered key to maximizing their contribution to food and nutritional security (Anon, 2023; Mayekiso, 2021; Shayanowako et al., 2021). This is most important in Africa, where hunger and malnutrition have attained a significant attention threshold (Gerrano, 2021; Zondi et al., 2022).

Scientists and policymakers are predicting that by 2050, food demand will increase by an estimated percent of about 70 to account for food and nutrition demands. Mayekiso et al. (2021) reported that despite the approaches adopted worldwide to mitigate food security issues in rural communities, poverty becomes a major constituent, translating to most rural households experiencing food insecurity shocks. South Africa is a developing country with a high unemployment rate (STATSSA, 2023), which triggers several households' susceptibility to food insecurity. The country is listed as one with diverse natural resources not limited to indigenous crops (vegetables and trees) (Shembe et al., 2023) that can be used as food by its citizens and, to some extent, businesses, especially in the rural areas, as these vegetables are native.

Report outputs by Alulu et al. (2023), Mayekiso (2021), and Shayanowako et al. (2021) confirmed the predominance of informal markets in the trading of indigenous leafy vegetables as they fail to penetrate formal markets because of poor product positioning and exclusion from produce demand and supply forecasts. The inception of cultivar development, mechanized processing methods, and market linkages will enhance the profitability of the indigenous leafy region. This review, therefore, would like to increase the insight into the state of different value chain components and assist with the upscaling and promoting the production of these crops, as well as inclusion in the current vegetable market and agricultural production. According to Shayanowako et al. (2021), there is an excellent opportunity for undeveloped and practical research in exploring indigenous leafy vegetables.

The review was compiled using an electronic search for peer-reviewed journals, conference materials, and international, regional, and national organizational reports published between 2016 and 2023 was used to conduct the literature search for this review. These online databases of the published literature identified were used to follow up on the current global trend new trends, which are required to solve current problems in the world's population nutritional status, viz., malnutrition and the existing trend to consume healthy foods, mainly of plant origin. African leafy vegetables (ALV), climate change, conservation techniques, genetic resources, crop mapping, and farmer's support are among the keywords utilized in the search for literature sources. Some of the terms were combined. The review mainly focused on indigenous (African) leafy vegetables, underutilized crops, and food and nutrition security.

1.1 Socio-Economic Study, Demand Assessment, Market, and Value Chain Analysis

Indigenous leafy vegetables (ILVs) are native to the' places where they are grown and are particularly well-adapted to the prevailing natural environmental conditions. Despite the significant role of indigenous vegetables on food security and their contribution to socio-economic surplus in many parts of the world, current observations in South Africa reveal that these crops have not been fully exploited. However, such vegetables can be cultivated locally as alternative food supplements and further be developed into different food produce, e.g., breakfast dishes similar to cornflakes and snacks to improve food production for many rural communities. This can increase communities income, promote the utilization of indigenous crops, improve the social well-being of communities, improve health status, and increase crop diversity. Limited knowledge and skills in cultivating and managing many of these indigenous species remain one of the significant production constraints. Hence, research needs to assess the potential and sustainability of these available indigenous vegetables to identify, recommend and explore various commercial product developments in the market. For countless South African families, especially in rural areas and townships, ILVs are a silent guardian, a crucial pillar of food security and nutrition. Imagine a mother with limited resources, knowing she can step into her garden or even nearby veld and gather a meal packed with essential vitamins and minerals for her children.

Odunitan et al. (2020) note that these leafy greens are "free or cheap" in most cases and offer vital micronutrients like Vitamins A and C, calcium, iron, and zinc. The issue is more than to stop hunger; it is to fight "hidden hunger"—the insidious absence of vital nutrients that can have a catastrophic effect on health and development. During droughts or economic stress, when staple foods don't perform, indigenous leafy vegetables (ILVs) are incredibly resilient, reflecting their evolutionary hardness.

Aside from the kitchen, these leafy greens are an essential asset. For many rural women, the market sale of excess indigenous leafy vegetables (ILVs) provides a valuable source of income. This income can be enough to buy school supplies, meet unexpected medical bills, or even contribute to the household's food rations. A study by Modi et al. (2006) has shown how important these small incomes are in strengthening household resilience. It is not a large business affair; instead, it is an essential component of the informal economy that holds people together.

Culturally, ILVs are integrated into who we are. They remind us of stories, traditions, and the folklore passed down over generations about how to grow, prepare, and even use them medicinally. If we forget these plants, then we risk forgetting part of ourselves, our collective knowledge, and our unique heritage.

Nonetheless, a shift is occurring. As urban development attracts individuals away from agrarian lifestyles and dietary patterns evolve toward more "contemporary" options, there exists a danger that indigenous local varieties (ILVs) are regarded as "food for the impoverished" (Modi et al., 2006). This viewpoint, combined with inadequate educational resources and opportunities in rural settings, can regrettably marginalize these remarkable plants from mainstream attention. While this presents a challenge, it simultaneously offers a chance to educate and motivate anew.

1.2 Crop-site Suitability Mapping

The vegetables (indigenous) were chosen based on their suitability to the area and nutritional value. Indigenous leafy vegetables are primarily collected from the wild, and little has been on their cultivation. Therefore, an attempt has to be made to identify suitable areas for massive crop production using remote sensing and GIS methodologies. The method can generate land cover data using SPOT 5 satellite imagery and modeling with the existing land capability. The research can further focus on suitable mapping areas for ALVs production using spatial modeling techniques and look at drought-tolerance and nutritional value of the crops.

The South African map below (Fig. 1) shows suitable areas for cultivating ILVs, where the deep greens would indicate areas of high suitability and where factors like rainfall, temperature, and soil pH create an ideal haven for growth. These might be regions known for consistent rainfall or specific soil types historically favored by these resilient plants. Conversely, lighter greens or yellows signify moderately suitable areas, perhaps requiring some management interventions like supplementary irrigation or soil amendments to optimize yields. However, orange or even reddish hues represent low suitability zones, areas where climatic conditions, soil composition, or water scarcity would make ILV cultivation particularly challenging or economically unfeasible.

Indigenous leafy vegetables have varying tolerances to heat and cold. For example, species like Amaranth (often called "morogo" in many parts of South Africa) are quite adaptable but still thrive best within specific

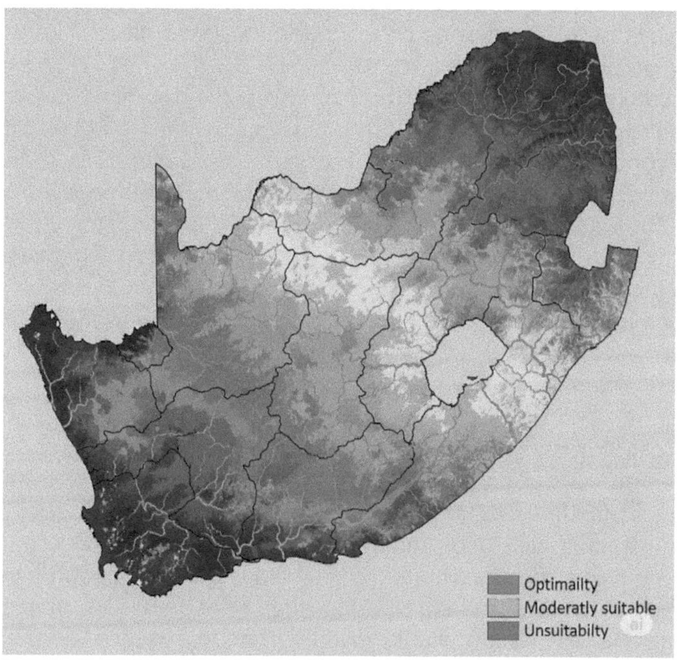

Fig. 1 Crop-site suitability map of indigenous leafy vegetable cultivation in South Africa (*Source* ARC)

thermal windows (Dube et al., 2012). In addition, many ILVs prefer well-drained soils, and their tolerance to acidity or alkalinity can vary significantly and water for consistent and successful cultivation, even for drought-tolerant varieties. Farmers can select specific ILV species best suited to their local conditions for optimum yield, leading to healthier plants and higher yields. This reduces the risk of crop failure, a significant concern for smallholder farmers. As Van Rensburg et al. (2007) noted, understanding environmental requirements is fundamental to improving the production of these underutilized crops. These vegetables can thrive, and the map actively encourages their cultivation and commercialization, helping to reverse the trend of their underutilization. This aligns with national efforts to diversify agricultural systems and leverage biodiversity for food security (DAFF, 2017).

1.3 Risk Assessment and Early Warning System

The importance of risk assessment and early warning systems is fundamental for managing droughts and pest control in ILV production. Risks can be detected through high resolution satellite imagery, which is becoming increasingly popular. The geographic information system (GIS)-based spatial analysis allows the project to identify localized hotspots of climate stress and pest outbreaks that threaten smallholder farmers and their production practices. Integrating GIS with early warning systems provides smallholder farmers with timely, site-specific alerts that support proactive decision-making and reduce production risks. Additionally, GIS-enabled risk mapping can improve smallholders' ability to optimize field management and enhance resilience across the ILV value chain. Implementing a climate smart production system is also vital, as it can be combined with various production techniques to maximize crop productivity and lessen the risk of crop failure during commercialization. Risk assessment is especially important for vulnerable smallholder farmers and communities facing threats to their water and land access, market access, information availability, and the overall vulnerabilities within food systems and nutrition value chains.

1.4 Genetic Resource Collection and Conservation

Genetic resource collection and maintenance are essential components of the commercialization stage of crops. It will play a vital role in meeting the future nutrition and food security of communities. The concept of germplasm conservation demands that collection methods initially capture maximum variation; subsequently, conservation and regeneration techniques minimize losses through time. The germplasm collection of ILVs will identify the diversity-rich areas of the world to enrich these genetic resources in South Africa. In the last decades, several studies have been conducted by the Agricultural Research Council (ARC) Genebank on the collection of ILV species and their variability among the germplasm in the world.

1.5 Pre-breeding and Breeding—Cultivar Development

The Pre-breeding and breeding of ILVs aim to reduce the genetic uniformity of the crops by using a different pool of genetic material to increase yield, improve nutritional value, and improve natural resource use efficiency and resistance to pests and diseases. Pre-breeding can be applied during the breeding programs, where desired traits are sought continuously and identified from source genotypes for use in cultivar development, identification of desirable crop traits and genes, and subsequent transfer into a suitable set of parents for further selection. The ARC is actively characterizing and evaluating ILVs germplasm to release new cultivars and to initiate breeding programs aimed at mitigating the effects of climate change is ongoing.

1.6 Agronomic Practices and Production System

The research by ARC focused on developing and optimizing cultivation practices. Many of the cultivation needs have now been identified, and research has shifted to more in-depth analysis, such as water use efficiency, nutritional water productivity, and the influence of climate change on crop yield and nutritional content. However, determining water use efficiency and nutritional water productivity of such crops is essential to saving water and nutrients for maximum farming profitability and ensuring food security and poverty alleviation for farmers and communities in South Africa. ARC has also investigated the productivity of indigenous leafy vegetables cultivated under dryland conditions. It is necessary to guarantee the quality of the product and gain the trust of consumers. Once the production process is optimized, the research will focus on the techno-economic of different production systems, such as hydroponics and shade net for growing ILVs.

1.7 *Development of Business Plans and Feasibility Study*

Indigenous leafy vegetables have a potential for value addition, such as in developing products in the cosmeceuticals and nutraceuticals industries. However, they have not been cultivated at a large scale before. Cultivating a small area and scaling up production over time, considering that this crop has not been grown successfully at a commercial level, is essential. In the first year of commercialization, market research will be required to determine the actual size of the market for ILVs and the demand as well. The study will also investigate other enterprise business models currently working on indigenous crops to look at best practices and lessons learnt.

1.8 *Training and Farmer Support*

On-farm demonstration sites will act as learning centers for farmers and extension officers through farmer's day and exhibitions (Fig. 1). The expected impact of this approach is that farmers participating in research will obtain firsthand knowledge of the ILV production practices, thereby enhancing the empowerment of communities (especially women, who have greater control of food provision in the families) and commercial and smallholder farmers. The transfer of technologies will minimize the impact of crop failure and create social cohesion by establishing social networks and motivation for community development, as well as reducing social alienation and family disintegration.

1.9 *Infrastructure Needs for Production and Processing*

Agro-processing technologies and product development of ILVs are limited. The competitive differentiation of drying solutions using a modular, fully integrated solar dehydration unit with a clean, standardized, and integrated processing unit for the food industries/specialized retail is essential. An innovative technology solution that combines the complete processing requirement (washing, blanching, and packaging) in one closed hygienic unit that can be either fixed infrastructure or mobile, can operate off the grid, requires low to no maintenance and is low in cost compared to available drying systems in the market is necessary.

Fig. 2 Training smallholder farmers on the cultivation of ILV and irrigation management (Chameleon soil water sensor)

Machinery such as mechanical harvester, slicing, packaging, milling, and radiating are required for agro-processing crops.

1.10 Post-Harvest Handling

Indigenous leafy vegetables are considerably more nutritious than their commercial counterparts, offering them the potential to combat malnutrition in underprivileged rural communities. However, further research on pre-and post-harvest management practices of leafy vegetables is required to improve crop productivity, storage, and shelf life while maintaining nutritional content (Fig. 2). No pest and disease management strategies are also currently employed during post-harvest handling (Fig. 3).

1.11 Marketing and Communication

Indigenous leafy vegetables are being neglected and consumed less in the modern commercial environment because they are not commercially farmed or distributed in the urban environment. They are primarily sold in the informal market (Table 1). These traditional crops have better

Fig. 3 Post-harvest technology: modified atmosphere packaging (MAP)

nutritional content than their modern substitutes, which currently leads to a detriment in nutrition for modern consumers. To bring these highly nutritious traditional vegetables back into modern diets, they need to be made easily accessible for consumers. This requires a new approach to its production and market route (which is currently limited to wild harvest and the informal local fresh markets). This includes a modernization of the value chain for indigenous crops, such as leveraging alternative uses for the vegetables as industrial and speciality ingredients, leveraging established supply chains as a superfood, and modernizing the element through quality assurance and technology to open new local and international markets. The target market for the final products of ALVs is dried food ingredient users, particularly those interested in indigenous crops and

Table 1 Return on investment of vendors from different African countries

Country	n	*Total volume (kg)*	*Average volume (kg)*	*Average price USD/kg*	*Total Turnover USD*	*Average Annual Turnover USD*
Benin	145	656,602	4,528	0.61	401,578	2769
Ivory Coast	140	99,877	713	0.54	53,544	382
Uganda	153	582,338	3,806	0.31	179,884	1176
Tanzania	179	1,986,760	11,099	0.23	45,789	2524
Senegal	143	1,654,474	11,570	1.47	243,7867	1,7048
South Africa	88	27,324	311	2.31	90,486	1028
Kenya	158	4,284,120	27,115	0.44	1,900,007	1,2025

nutritional value, like health and food supplement houses, processed food industry and large foodservice operators, especially those with affinity to traditional values. The ARC is also collaborating with food companies that are interested in developing products from indigenous crops and communities that can be linked to produce from them to generate income. For instance, the market for dried fruit and vegetables is worth USD 40 billion annually in the EU alone. Of this, the majority are dried fruits and herbs, which ALVs have the potential to share in the market as superfoods.

2 Conclusion

The edible parts of indigenous leafy vegetables like Amaranth, Cowpea, Corchorus, and others, such as leaves and grains, contain proteins, fiber, squalene, polyphenols, and flavonoids, among others, which present different biological activities with a positive impact on health, and which have allowed the formulation of novel functional foods. Despite the potential for these vegetables to take over other imported vegetables and become a competitive source of food worldwide, due to their ability to suit ideally organic farming and contribute to climate impact mitigation strategies, a great deal of scientific-technological effort is required, such as breeding and DNA fingerprinting. From a technological point of view, work should be done on the development of new functional foods with proven biological effects.

Acknowledgements The authors would also like to thank the Agricultural Research Council and the Department of Agriculture, Land Reform, and Rural Development for the financial support of the study.

References

Anon, M. C. (2023). Health benefits of Amaranth. *Sustainable Food Science: A Comprehensive Approach, 2*. https://doi.org/10.1016/B978-0-12-823960-5.00010-X

Alulu, J., Makyao, M., Huyskens-keil, S., Lenz, B., Muendo, K. M., Mganilwa, Z., Mbeche, R., Mgaya, P., & Mithöfer, D. (2023). Distribution, transportation, and coordination in African indigenous vegetable value chains: A scoping review. *Frontiers in Environmental Economics, 2*, 1113826. https://doi.org/10.3389/frevc.2023.1113826

Azam-Ali, S., Ahmadzai, H., Choudhury, D., Goh, E. V., Jahanshiri, E., Mabhaudhi, T., Meschinelli, A., Modi, A. T., Nhamo, N., & Olutayo, A. (2023). Marginal areas and indigenous people priorities for research and action. In J.von Braun et al. (Eds.), *Science and innovations for food systems transformation*. https://doi.org/10.1007/978-3-031-15703-5_14

Bokelmann, W., Huyskens-Keil, S., Ferenczi, Z., & Stöber, S. (2022). The role of indigenous vegetables to improve food and nutrition security: Experiences from the Project HORTINLEA in Kenya (2014–2018). *Frontiers in Sustainable Food Systems, 6*, Article 806420. https://doi.org/10.3389/fsufs.2022.806420

Chanza, N., & Musakwa, W. (2022). Revitalizing indigenous ways of maintaining food security in a changing climate: A review of the evidence base from Africa. *International Journal of Climate Change Strategies and Management, 14*(3), 252–271.

Department of Trade, Industry and Cooperation (DTIC). (2023). Opinion piece: Africa agricultural intra-trade to advance food security and industrial development—challenges and opportunities. Chief Director: Makhwasane Andrew Matjeke. Economic Research and Coordination (ERC) Branch, Macroeconomic Policy Unit, Pretoria, South Africa.

DAFF (Department of Agriculture, Forestry and Fisheries). (2017). National Policy on Food and Nutrition Security for South Africa. Pretoria: Government Printers (While not directly a map citation, this policy document underpins the importance of diverse food systems, including ILVs).

Dube, T., Modi, A. T., & Mabhaudhi, T. (2012). The potential of selected indigenous African leafy vegetables in alleviating food insecurity in South Africa. *African Journal of Agricultural Research, 7*(30), 4210–4219.

Gerrano, A. (2021). Indigenous African Vegetables can contribute to food and nutritional security. Integrated farm management: Integrated pest management, rethink crop production, nurturing soil's metabolic potential. *Agri-About, 101.*

Mayekiso, A. (2021). Economic assessment of indigenous leafy vegetables (ILVs) production for income generation and food security in the Eastern Cape province, South Africa. Doctor of Philosophy in Agriculture (Agricultural Economics) Thesis, Department of Agricultural Economics and Animal Production, Faculty of Science and Agriculture, School of Agricultural and Environmental Sciences, University of Limpopo, Polokwane, South Africa.

Mabhaudhi, T., Chibarabada, T., Chimonyo, V., Murugani, V., Pereira, L., Sobratee, N., Govender, L., Slotow, R., & Modi, A. (2019). Mainstreaming underutilized indigenous and traditional crops into food systems: A South African perspective. *Sustainability, 11*(1). https://doi.org/10.3390/su11010172

Mabhaudhi, T. V., Chimonyo, G. P., & Modi, A. T. (2017). Status of underutilised crops in South Africa: Opportunities for developing research capacity review. *Sustainability, 9*, 1–21.

Mudau, F. N., Chimonyo, V. G. P., Modi, A. T., & Mabhaudhi, T. (2022). Neglected and underutilized crops: A systematic review of their potential as food and herbal medicinal crops in South Africa. *Frontiers in Pharmacology, 12*, 1–31.

Modi, A. T., & Mabhaudhi, T. (2016). Developing a research agenda for promoting underutilised, indigenous and traditional crops. Crop Science School of Agricultural, Earth and Environmental Sciences, University of KwaZulu-Natal. Water Research Commission Report No. WRC Report No. KV 362/16. ISBN 978-1-04312-0891-3.

Modi, M., Mabhaudhi, T., & Odindo, A. (2006). African indigenous vegetables: An untapped food and nutritional resource for South Africa. *South African Journal of Science, 102*(11/12), 478–482.

Nhamo, L., Paterson, G., Van Der Walt, M., Moeletsi, M., Modi, A., Kunz, R., Chimonyo, V., Masupha, T., Mpandeli, S., Liphadzi, S., Molwantwa, J., & Mabhaudhi, T. (2022). Optimal production areas of underutilized indigenous crops and their role under climate change: Focus on Bambara groundnut. *Frontiers in Sustainable Food Systems, 6*, 990213. https://doi.org/10.3389/fsufs.2022.990213

Omotato, A. O., & Aremu, A. O. (2020). Underutilized African indigenous fruit trees and food–nutrition security: Opportunities, challenges, and prospects—Review. *WILEY—Food and Energy Security,* 1–16. https://doi.org/10.1002/fes3.220

Odunitan, O. A., Ngoma, L., & Ngxwala, N. (2020). Socio-economic benefits and market access for indigenous leafy vegetables in South Africa: A systematic review. *Journal of Agricultural Science, 12*(7), 108–117.

Qwabe, Q. N., & Pittawaty, T. (2023). Exploring the role of indigenous vegetables in rural livelihoods: Perceptions from the Ntuze Community. *South African Journal of Agricultural Extension (SAJAE), 51*(1), 142–154. https://doi.org/10.17159/2413-3221/2023/v51n1a14377

Shayanowako, A. I. T., Morrissey, O., Tanzi, A., Muchuweti, M., Mendiondo, G. M., Mayes, S., Modi, A. T., & Mabhaudhi, T. (2021). African leafy vegetables for improved human nutrition and food system resilience in Southern Africa: A scoping review. *Sustainability, 13*, 2896.

Shembe, P. S., Ngobese, Z. N., Siwela, M., & Kolanisi, U. (2023). The potential repositioning of South African underutilised plants for food and nutrition security: A scoping review. *Heliyon, 9*, 1–13.

Van Rensburg, W. J., Van Averbeke, W., & Slabbert, E. (2007). African leafy vegetables in South Africa: A review of their cultivation, harvesting, processing and consumption. Pretoria: Water Research Commission (This type of review would often contain information on environmental needs for various ILVs).

Zondi, N. T. B., Ngidi, M. S. C., Ojo, T. O., & Hlatshwayo, S. I. (2022). Impact of market participation of indigenous crops on household food security of smallholder farmers of South Africa. *Sustainability, 14*, 15194.

CHAPTER 3

Evaluating the Potential of Agrivoltaic Systems for Mitigating Climate Change Impacts in BRICS Countries

Ravi Kumar Chopdar and Namrata Sengar

1 Introduction

Climate change is a major threat to agricultural systems worldwide, including in the BRICS (Brazil, Russia, India, China, and South Africa) nations. Due to their large populations, extensive agricultural sectors, and diverse climatic conditions, these nations are particularly vulnerable to climate change's effects on food security, water resources, and rural livelihoods. The BRICS nations have a substantial economic and demographic impact. They occupy 30% of the world's land, house 40% of the world's population, and contribute over 50% of the world's GDP (Abbas et al., 2022; Liu et al., 2022).

R. K. Chopdar (✉) · N. Sengar
Department of Pure and Applied Physics, University of Kota, Kota, Rajasthan, India
e-mail: chopdarravi92@gmail.com

N. Tshikovhi et al. (eds.), *The BRICS Imperative*,
https://doi.org/10.1007/978-3-032-05932-1_3

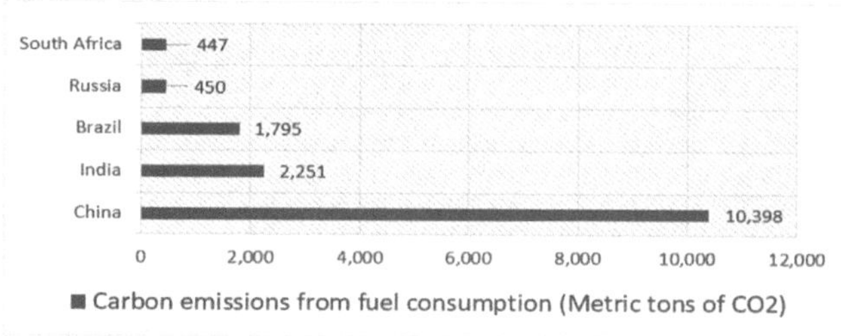

Fig. 1 Carbon emission from fuel consumption in BRICS (*CO2 Emissions from Fuel Combustion | World Statistics on CO2 Updated | Enerdata*, n.d.)

Rapid economic growth and population growth in the BRICS countries have resulted in increased fossil fuel use and greenhouse gas emissions. BRICS nations consume 36% of global energy (Hasan, 2022). Climate change is caused by energy sector CO_2 emissions. China, India, and Russia emit the most greenhouse gases, followed by Brazil and South Africa, as shown in Fig. 1. To limit the rise in global temperatures and combat climate change, it is essential to reduce emissions.

The BRICS countries have a big impact on the world's population, economy, and environment. They are actively converting to renewable energy sources and implementing green energy technology in response to increasing emissions and climate issues (Abbas et al., 2022; Miranda et al., 2021).

Solar energy is clean, reliable, and sustainable. It is more stable, cost-effective, and affordable due to technological advancements. It generates electricity without releasing any greenhouse gases or pollutants into the atmosphere. Energy independence, new employment opportunities, and a more sustainable future are all benefits of solar power (Dupraz et al., 2011). Solar power generation is expanding rapidly and might fulfill a large amount of the world's energy needs (REN21, 2021). However, ground-based solar panels can put pressure on agricultural land, raising questions about the use of land for both electricity generation and food production. Integrating solar power plants on agricultural land can improve land utilization and eliminate conflicts between food and energy (Dos Santos, 2019). This strategy provides an opportunity to resolve the amount of land required for large-scale solar farm installations and reduce the conflict between food and energy needs.

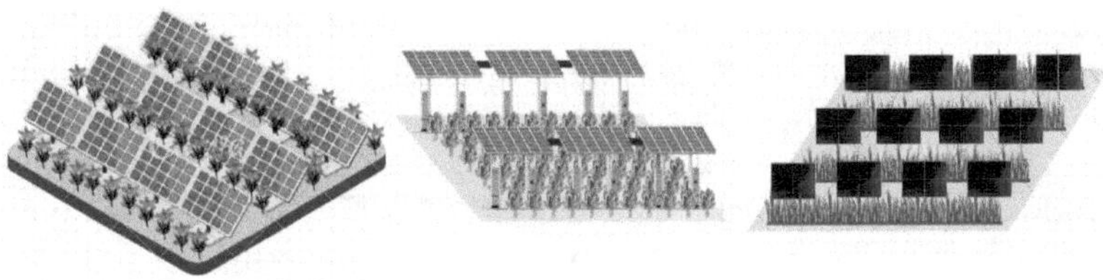

Fig. 2 Different types of AVS

Agrivoltaic system (AVS) combines the production of photovoltaic (PV) power and farming on the same piece of land. This method involves putting PV panels on the ground and growing crops in the areas between them. Also, if the PV array is built high enough, the space under the PV panels can be used for farming (Chopdar et al., 2024, 2025; Poonia et al., 2022). Figure 2 depicts different types of AVS.

This paper aims to evaluate the potential of agrivoltaic (AV) systems for mitigating climate change impacts in the BRICS countries comprehensively. BRICS solar potential will be examined. Agrivoltaics is a viable solution due to renewable energy generation, land utilization, climate resilience, water conservation, economic benefits, policy support, sociocultural acceptance, and the need for collaboration. We'll also examine each BRICS country's climate and AVS projects. Integrating renewable energy and agriculture to combat climate change requires understanding BRICS AVS potential. Policymakers can improve sustainability, water management, emissions, and livelihoods by evaluating feasibility, benefits, and inter-disciplinary approaches. Agrivoltaics provides food and energy security for growing BRICS populations by combining agriculture and solar energy.

2 Solar Potential in BRICS Countries

Solar resource maps (Figs. 3, 4, 5, 6, 7, 8, 9, 10, 11, and 12) show that the BRICS nations of China, India, Brazil, and South Africa all have substantial potential for harnessing solar radiation. Russia has potential to use solar energy as a clean, renewable power source despite its lower solar potential. Figure 3 shows that China has a global horizontal irradiation (GHI) range of 4 to 5 kWh/m^2, with slightly lower levels found in certain middle regions such as Chengdu and Chongqing, and some western regions with irradiation rates exceeding 5 kWh/m^2. According to Fig. 4, the daily total photovoltaic (PV) power output for China as a whole, ranges between 4 and 6 kWh/kWp. The output ranges from 1400 to 2000 kWh/kWp annually. The majority of India has a GHI between 5 and 6 kWh/m^2 (Fig. 5), with the exception of some northern and north-eastern regions. The daily PV output potential ranges between 4 and 5 kWh/kWp, while the annual potential ranges between 1,400 and 1,900 kWh/kWp, as shown in Fig. 6. Figure 7 shows that the GHI levels for South Africa range from 4.2 to 6.2 kWh/m^2. Figure 8 shows that the PV power output potential in South Africa is between 4 and 5.4 kWh/kWp per day and between 1400 and 2000 kWh/kWp per year.

Figure 9 shows that the GHI in Brazil ranges from 5 to 6.2 kWh/m^2 over a large part of the country. Figure 10 shows that the total PV power output potential in Brazil is between 4 and 4.8 kWh/kWp per day, and between 1400 and 1800 kWh/kWp per year. Figure 11 shows that the world horizontal irradiation values for the Russian Federation range from

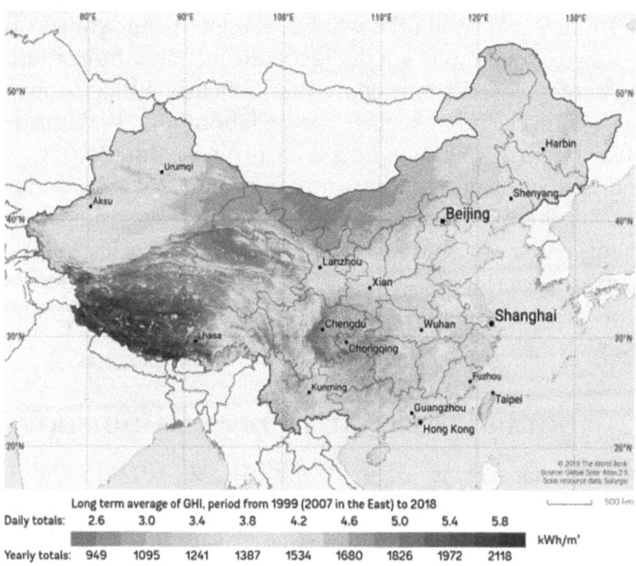

Fig. 3 Solar resource map for global horizontal irradiation for China (Solargis)

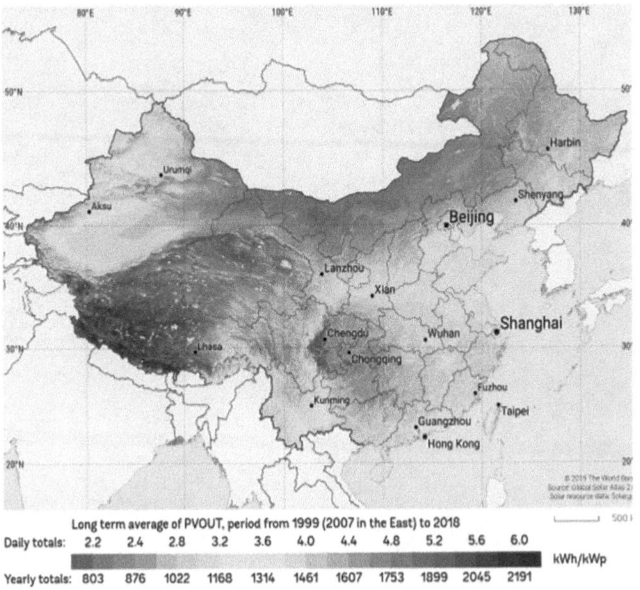

Fig. 4 Solar resource map for PV power potential for China (Solargis)

2.8 to 4.6 kWh/m^2. Figure 12 shows that the daily solar output potential in Russia is between 3 and 5 kWh/kWp, while the average annual output is between 1000 and 1800 kWh/kWp. Table 1 summarizes the GHI and PVout values for BRICS countries.

BRICS countries can adopt agrivoltaic (AV) systems to improve renewable energy generation because of their higher solar potential. This minimizes greenhouse gas emissions as well as their reliance on fossil fuels. AV systems provide additional benefit such as increased land use efficiency, which can mitigate the conflict between food and energy. These countries contribute to a greener and more ecologically friendly energy sector by supporting sustainable farming methods.

3 Current Status of AVS in BRICS

This section discusses the current state of AVS in the BRICS countries.

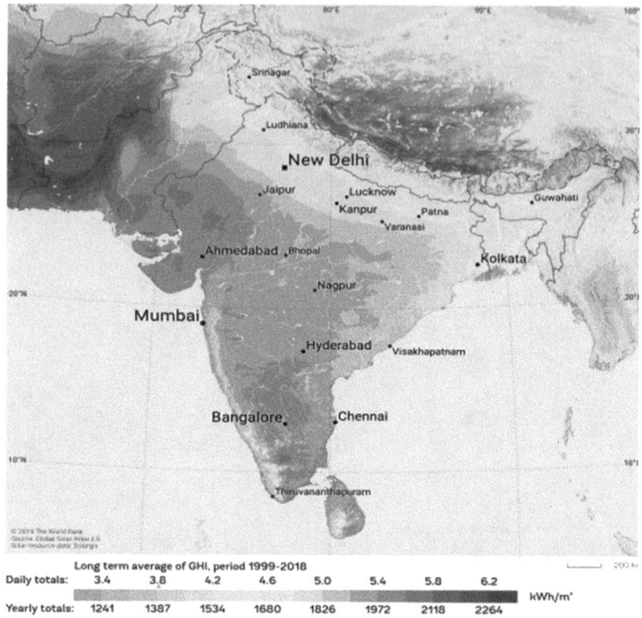

Fig. 5 Solar resource map for global horizontal irradiation for India (Solargis)

3.1 *Brazil*

The first AVS in Brazil, named Ecolume, was created through a collaborative effort involving a network of over 40 researchers from Brazil in 2019. The project received funding from CNPq (National Council for Scientific and Technological Development). The project for Ecolume was installed at the SERTA agroecology school in Ibimirim, a municipality located in the state of Pernambuco (Lacerda et al., 2020). This plant has shown notable benefits for food production, particularly in temperate region (Nobre et al., 2019).

3.2 *Russia*

While Russia as a whole generates only 0.2% of its electricity from wind and solar sources (Dave Jones et al., 2020), certain regions, including the Northern Caucasus and Krasnodar region, have sufficient solar potential for agrivoltaic installations (Kostik et al., 2020). This is supported by a

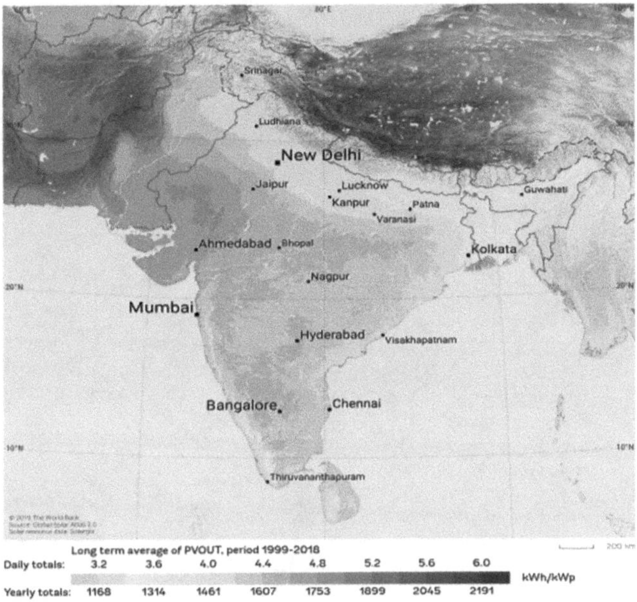

Fig. 6 Solar resource map for PV power potential for India (Solargis)

study that focuses on the southern region of Russia and highlights the high solar activity and availability of agricultural land in these areas. The findings show that with the right slope and spacing, large amounts of electricity can be generated in these areas using agrivoltaics (Kostik et al., 2020).

3.3 *India*

Researchers from International Institute for Sustainable Development published a report that profiles fourteen active installations across India, providing details on their locations, sizes, crops, and panel designs (Rahman et al., 2023). In Maharashtra, Gro Solar Energy Private Limited has launched a 7 MW solar project that incorporates low-height crops. There are four large commercial hybrid plants with a combined capacity of 1054 kWp in the Indian state of Gujrat. The Central Arid Zone Research Institute in Jodhpur, Rajasthan (2017) and Jalgaon, Maharashtra (2014)

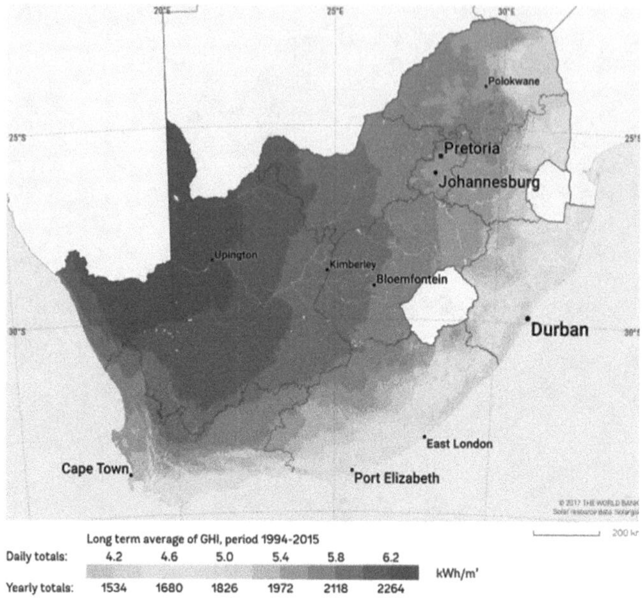

Fig. 7 Solar resource map for global horizontal irradiation for South Africa (Solargis)

both have AV systems that are used for research (Poonia et al., 2022; Rahman et al., 2023).

3.4 China

China has the world's largest agrivoltaic system. This extraordinary Baofeng Group project covers 20 million square meters in Ningxia's desert. The Baofeng Group and Huawei partnership has promoted renewable energy and sustainable agriculture in China (*Giant Agrivoltaic Project in China | Distretto Atena Future Technology*, n.d.; Market Research Study Agrivoltaics, n.d.). In 2016, REM TEC built an agrivoltaic plant in the county of Jinzhai in the province of Anhui. In 2016, Panda Green Energy put solar panels on top of farms in Turpan. After an audit in October 2017, the company got permission to use its method all over the country. There have been projects built on a MW size.

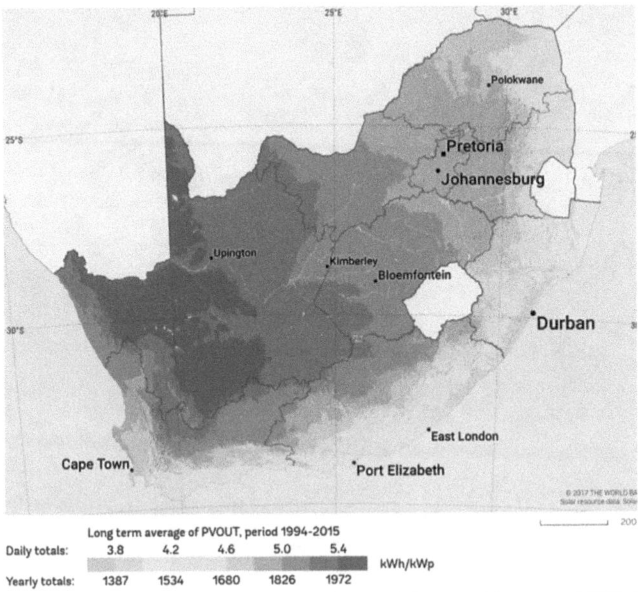

Fig. 8 Solar resource map for PV power potential for South Africa (Solargis)

In 2016, the province of Jiangxi built a 70 MW agrivoltaic plant. In 2017, Fuyang Angkefeng Optoelectronic Technology built a 50 KWp test site for agrivoltaics in Fuyang, Anhui Province. The Baofeng Group is building a 1GW AVS in Binhe New District with goji berry plants, Ningxia Province, on the eastern sides of the Yellow River. Huawei's inverters have connected 640 megawatts of the solar farm to the grid (*Giant Agrivoltaic Project in China | Distretto Atena Future Technology*, n.d.). China had an impressive solar capacity of about 1.9 gigawatts in 2020. Solar greenhouses have also transformed the country's agriculture. Solar greenhouses have cut energy use and CO2 emissions north of the Huai River and surrounding Beijing (Market Research Study Agrivoltaics, n.d.). AV systems allow farmers to generate electricity on their farms and become energy independent. This can increase the agricultural sector's utilization of renewable energy. In China, promoting agrivoltaics is consistent with emission reduction objectives and supports the country's climate commitment (Liu, 2020).

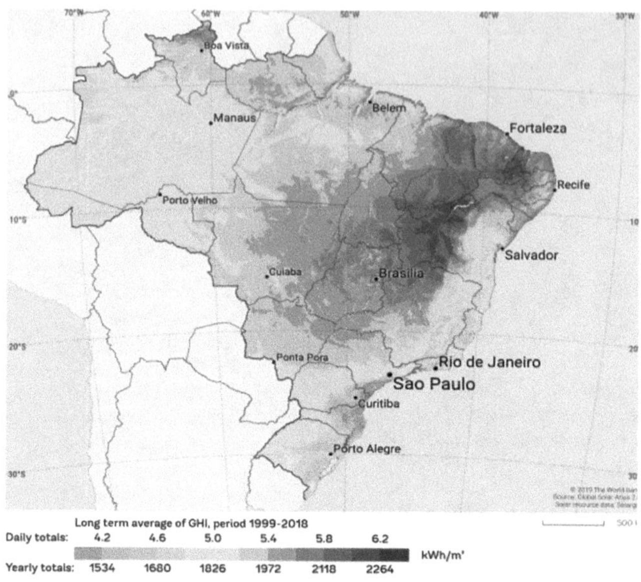

Fig. 9 Solar resource map for global horizontal irradiation for Brazil (Solargis)

3.5 South Africa

In South Africa, only 10% of all the projects were solar PV installations on farms. Even though solar photovoltaics had been made available, they were still in their early stages of use. The Eskom pilot project at the Komati power station was a good example of large-scale implementation because it used land that was already there to build an agrivoltaic plant. South Africa has taken steps toward using solar energy in farming by putting out requests for proposals (*Understanding What Is Solar Agrivoltaics & Applications for Southern Africa—Zero Point Energy*, n.d.).

BRICS countries have tried different approaches and projects related to AVS to meet energy needs and encourage sustainable practices.

4 Potential of AVS in BRICS

Countries with great solar potential, such as Brazil, India, China, and South Africa, can benefit from AV systems in a variety of ways. These systems make a fair amount of electricity possible through high solar

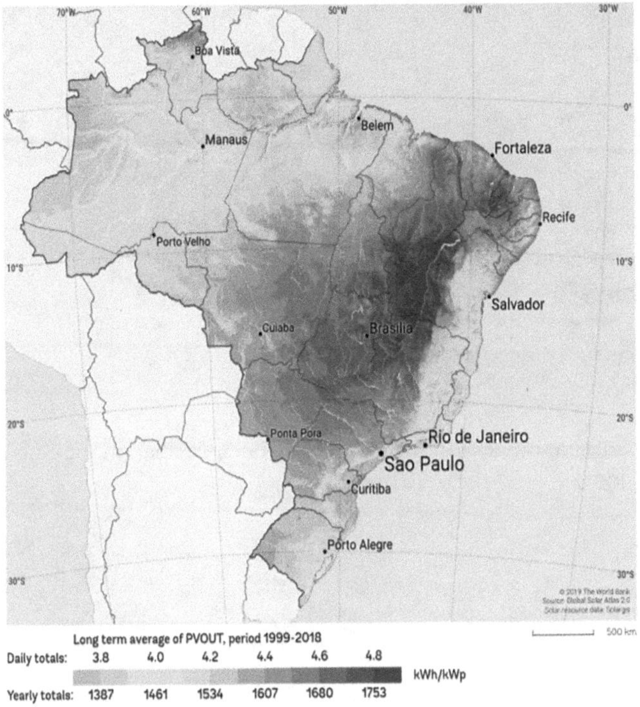

Fig. 10 Solar resource map for PV power potential for Brazil (Solargis)

potential, which leads to more green energy production. Integrating solar panels makes the best use of space, allowing energy to be made and crops to be grown at the same time. AV systems also help reduce greenhouse gas emissions and the reliance on fossil fuels. This makes farming more sustainable and makes the energy industry greener.

India is second only to China in terms of development in the renewable energy sector. Solar photovoltaic and wind energy account for more than three-quarters of capacity additions. Currently, renewable energy sources account for 9.11% (10,3 billion units) of India's total power generation (113,2 billion units), with solar power accounting for 36.3%. India's renewable energy potential is expected to increase from 131 GW in 2019 to 275 GW in 2027 (Poonia et al., 2022). While Russia as a whole has

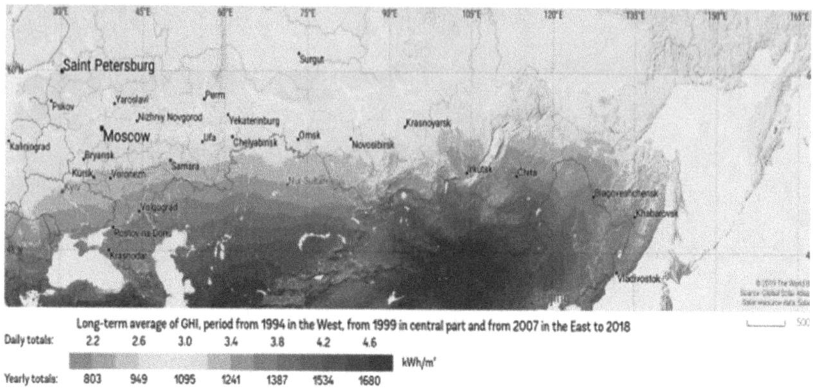

Fig. 11 Solar resource map for global horizontal irradiation for Russian Federation (Solargis)

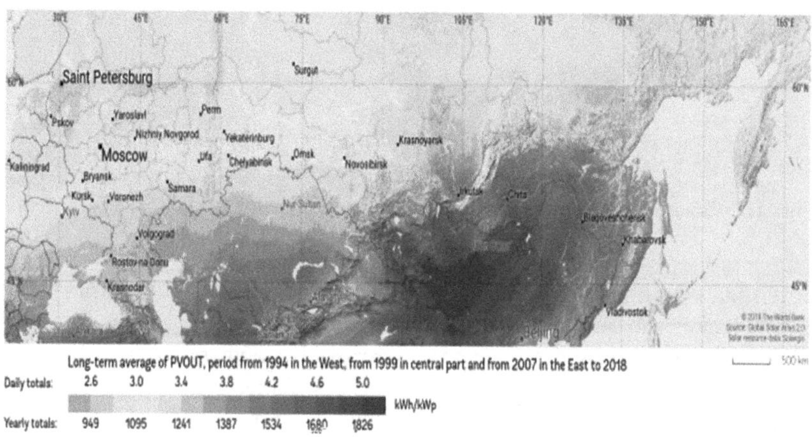

Fig. 12 Solar resource map for PV power potential for Russian Federation (Solargis)

a lower solar potential, its southern regions have sufficient solar potential to support the installation of AVS. By employing this technology, Russia can combat climate change and contribute to the sustainability

Table 1 GHI and PVout (daily and annual basis) values for BRICS countries

Country	*GHI (kWh/m²)*	*Daily PV Power Output Potential (kWh/kWp)*	*Yearly PV Power Output Potential (kWh/kWp)*
Brazil	5–6.2	4–4.8	1400–1800
Russia	2.8–4.6	3–5	1000–1800
India	5–6	4–5	1400–1900
China	4–5	4–6	1400–2000
South Africa	4.2–6.2	4–5.4	1400–2000

of agriculture in its climate (Kostik et al., 2020). The climate and geography of Southern Africa make agrivoltaics a promising field of research and development. The region has a lot of farmlands because of the favorable climatic conditions and terrain, which allow for a variety of large-scale agricultural production. According to World Bank's latest data, 79.4% of South Africa is farmland (*Agricultural Land (% of Land Area)—South Africa | Data*, n.d.). Solar PV systems would be set up on current farmland, making efficient use of the available space (*Understanding What Is Solar Agrivoltaics & Applications for Southern Africa—Zero Point Energy*, n.d.). AV systems combine farming with solar energy to create a controlled environment that helps crops grow better and saves water.

BRICS countries with significant solar potential, such as Brazil, India, China, and South Africa, can benefit from AV systems by producing renewable energy and promoting sustainable agriculture. India and Russia have the potential to develop renewable energy despite their varying solar potential. Adopting agrivoltaics can result in a greener future for these nations by combining energy generation with food production and combating global warming.

5 Result and Discussion

The BRICS countries have significant solar potential, with favorable GHI values ranging from 4 to 6.2 kWh/m^2. This makes them ideal candidates for harnessing solar radiation to meet the increasing energy demands caused by population growth, urbanization, and industrialization. Concurrently, these nations are major contributors to global food production, necessitating the implementation of sustainable agricultural practices. By combining solar energy generation with agriculture, AV systems offer a promising solution for meeting both energy and food needs in a sustainable manner.

South Africa and Brazil, with their abundant solar potential and arid climate, are ideal locations for AV systems. These countries can considerably benefit from the installation of agrivoltaics, which can help mitigate the effects of climate change, increase land use efficiency, and contribute to sustainable development. In Pernambuco, Brazil, the Ecolume project has successfully integrated AVS to increase food production in temperate regions (Chiengkul, n.d.). Even though Russia has a lower overall solar potential, certain regions, such as the Northern Caucasus and Krasnodar, have sufficient solar activity for photovoltaic installations (Kostik et al., 2020). As the third-largest emitter of greenhouse gases, India has significant sources of emissions in the electricity, agriculture, manufacturing, and construction industries. Implementing agrivoltaics can aid in the production of additional electricity, thereby reducing reliance on coal-based power facilities. In addition, agrivoltaics can provide a new source of revenue, decreasing the need for livestock in farming and reducing the release of the potent greenhouse gas methane. India and Germany have collaborated to profile active AVS installations throughout the country, including commercial hybrid plants and research facilities in Maharashtra, Gujarat, and Rajasthan (Rahman et al., 2023). China, the largest emitter of greenhouse gases, has significant solar energy potential. By leveraging solar energy for both food production and electricity generation, AV systems can play a vital role in China's climate change mitigation efforts. Agrivoltaics can help China meet its food security and climate change mitigation goals, which are especially important given the country's massive population. In the desert of Ningxia, China has the largest agrivoltaic project in the world, demonstrating the potential for renewable energy and sustainable agriculture. South Africa has demonstrated an interest in solar energy for agriculture through the Eskom pilot project at the Komati power station.

BRICS countries have substantial potential for AV systems. Particularly China and India can benefit from agrivoltaics by mitigating the effects of climate change and achieving food security and energy sustainability. Figure 13 depicts the environmental benefits of AVS, includes reduced land use, reduced GHG emissions, water conservation, and improved crop microclimate. Implementing agrivoltaics as a comprehensive solution can be a game-changer in BRICS countries' efforts to combat climate change and promote sustainable agriculture.

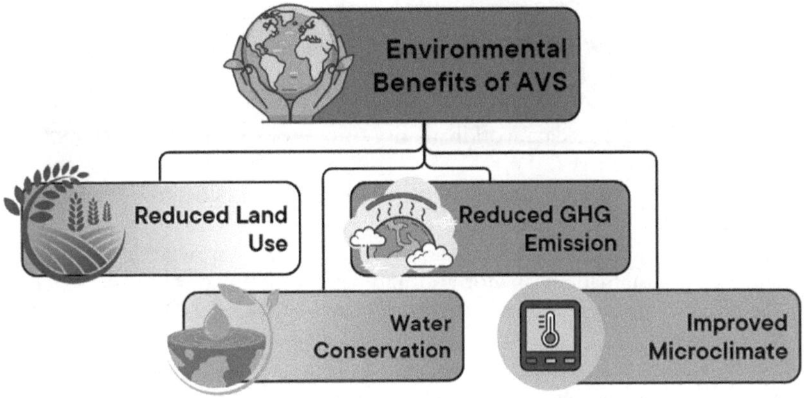

Fig. 13 Environmental benefits of AVS

6 Conclusion

AV systems are a promising strategy for addressing the effects of climate change in the BRICS nations. Brazil, India, China, and South Africa can benefit from AVS by generating renewable energy and promoting sustainable agriculture due to their substantial solar potential. Integrating solar panels with agricultural land optimizes space utilization and reduces competition between food production and energy production. These nations can reduce greenhouse gas emissions, improve energy security, and contribute to a greener and more sustainable future by adopting agrivoltaics. The successful implementation of AVS in various regions of the BRICS nations, such as the Ecolume project in Brazil and large-scale projects in China, demonstrates the potential for wider adoption. Policymakers can play a crucial role in combating climate change and achieving a more resilient, low-carbon future by supporting AVS development through the implementation of farmers-supporting policies and the collaboration of other stakeholders.

References

Abbas, S., Gui, P., Chen, A., & Ali, N. (2022). The effect of renewable energy development, market regulation, and environmental innovation on CO2 emissions in BRICS countries. *Environmental Science and Pollution Research, 29*(39), 59483–59501. https://doi.org/10.1007/S11356-022-20013-7/METRICS

Agricultural land (% of land area)—South Africa | Data. (n.d.). Retrieved July 16, 2023, from https://data.worldbank.org/indicator/AG.LND.AGRI.ZS?locations=ZA&view=chart

Chiengkul, P. (n.d.). Agrivoltaic systems and just energy-agriculture transition in Southeast Asia. 29, 2023.

Chopdar, R. K., Sengar, N., & Giri, N. C. (2025). Agrivoltaic systems for enhancing sustainability in hot arid and semi-arid climates. *Lecture Notes in Electrical Engineering, 1314*, 107–119. https://doi.org/10.1007/978-981-96-0861-4_8

Chopdar, R. K., Sengar, N., Giri, N. C., & Halliday, D. (2024). Comprehensive review on agrivoltaics with technical, environmental and societal insights. *Renewable and Sustainable Energy Reviews, 197*, Article 114416. https://doi.org/10.1016/j.rser.2024.114416

CO2 Emissions from fuel Combustion | World Statistics on CO2 Updated | Enerdata. (n.d.). Retrieved July 6, 2023, from https://yearbook.enerdata.net/co2/emissions-co2-data-from-fuel-combustion.html

Jones, D., Graham, E., & Tunbridge, P. (2020). *Wind and solar now generate one-tenth of global electricity global half-year electricity analysis.*

Dos Santos, C. (2019). *Agrivoltaic system: a possible synergy between agriculture and solar energy.* https://www.diva-portal.org/smash/get/diva2:1427912/FULLTEXT01.pdf

Dupraz, C., Marrou, H., Talbot, G., Dufour, L., Nogier, A., & Ferard, Y. (2011). Combining solar photovoltaic panels and food crops for optimising land use: Towards new agrivoltaic schemes. *Renewable Energy, 36*(10), 2725–2732. https://doi.org/10.1016/j.renene.2011.03.005

Giant agrivoltaic project in China | Distretto Atena Future Technology. (n.d.). Retrieved July 12, 2023, from https://www.atenaweb.com/giant-agrivoltaic-project-in-china/

Hasan, M. (2022). Energy economic expansion with production and consumption in BRICS countries. *Energy Strategy Reviews, 44*, Article 101005. https://doi.org/10.1016/J.ESR.2022.101005

Kostik, N., Bobyl, A., Rud, V., & Salamov, I. (2020). The potential of agrivoltaic systems in the conditions of southern regions of Russian Federation. *IOP Conference Series: Earth and Environmental Science, 578*(1), Article 012047. https://doi.org/10.1088/1755-1315/578/1/012047

Lacerda, F. F., Lopes, G. M. B., Coutinho, R. D. da S., Dos Santos, S. A., Da Silva, M. V., Sabino, H. B., & Lima, J. P. V. de. (2020). O Projeto Ecolume: O paradigma da abundância na convivência com o clima semiárido no Nordeste brasileiro. *Revista Fitos, 14*(2), 207–221. https://doi.org/10.32712/2446-4775.2020.941

Liu, F., Zhang, X., Adebayo, T. S., & Awosusi, A. A. (2022). Asymmetric and moderating role of industrialisation and technological innovation on energy intensity: Evidence from BRICS economies. *Renewable Energy, 198*, 1364–1372. https://doi.org/10.1016/J.RENENE.2022.08.099

Liu, Y. (2020). *Farming the sun and the crops at once: A cost benefit-analysis of implementing an agrivoltaic system in China*. Nature.Berkeley.Edu. https://nature.berkeley.edu/classes/es196/projects/2020final/LiuY_2020.pdf

Market Research Study Agrivoltaics. (n.d.). Retrieved July 12, 2023, from https://science.osti.gov/-/media/sbir/pdf/Market-Research/SETO---Agrivoltaics-August-2022-Public.pdf

Miranda, I. T. P., Moletta, J., Pedroso, B., Pilatti, L. A., & Picinin, C. T. (2021). A review on green technology practices at BRICS Countries: Brazil, Russia, India, China, and South Africa. *SAGE Open, 11*(2). https://doi.org/10.1177/21582440211013780/ASSET/IMAGES/LARGE/10.1177_21582440211013780-FIG6.JPEG

Nobre, P., Pereira, E. B., Lacerda, F. F., Bursztyn, M., Haddad, E. A., & Ley, D. (2019). Solar smart grid as a path to economic inclusion and adaptation to climate change in the Brazilian Semiarid Northeast. *International Journal of Climate Change Strategies and Management, 11*(4), 499–517. https://doi.org/10.1108/IJCCSM-09-2018-0067/FULL/PDF

Poonia, S., Jat, N. K., Santra, P., Singh, A. K., Jain, D., & Meena, H. M. (2022). Techno-economic evaluation of different agri-voltaic designs for the hot arid ecosystem India. *Renewable Energy, 184*, 149–163. https://doi.org/10.1016/J.RENENE.2021.11.074

Rahman A., Sharma A., Postel F., Goel S., Kumar K., & Laan T. (2023). *Agrivoltaics in India Challenges and opportunities for scale-up IISD REPORT International Institute for Sustainable Development*. https://policycommons.net/artifacts/3867686/agrivoltaics-in-india/4673913/

REN21. (2021). *REN21: Renewables 2017 Global Status Report*. https://scholar.google.com/scholar?hl=en&as_sdt=0%2C5&q=REN21.+Renewables+2017+global+status+report%2C+Paris%3A+REN21+Secretariat.+ISBN+978-+3-9818107-6-9&btnG=

Understanding What is Solar Agrivoltaics & Applications for Southern Africa - Zero Point Energy. (n.d.). Retrieved July 15, 2023, from https://zpenergy.co.za/understanding-what-is-solar-agrivoltaics-applications-for-southern-africa/

CHAPTER 4

Higher Education for Sustainable Development: Brazil and South Africa's Path Trajectory Towards SDG 4

Carlo Emile Daniels

1 Introduction and Background

Although countries worldwide have made significant strides in their efforts to increase educational access, approximately 260 million children were out of school in 2018 (UNICEF, 2021). Despite the efforts and dedication of governments, educational leaders, and educators to improve the quality of education, more than half of all children and adolescents globally are struggling to meet minimum proficiency standards in reading and mathematics (UNICEF, 2021). According to the Civil Society Working Group for the 2030 Agenda (2020: 4) and Friedman (2020), the emergence of Covid-19 has exacerbated the aforesaid challenges and has revealed the true nature of inequality in South Africa and Brazil. As reported by The Human Rights Watch website (2021),

C. E. Daniels (✉)
Department of Educational Foundations, University of South Africa, Pretoria, South Africa
e-mail: daniece@unisa.ac.za

N. Tshikovhi et al. (eds.), *The BRICS Imperative*,
https://doi.org/10.1007/978-3-032-05932-1_4

in November 2020, five million Brazilian children (between the ages of 6 and 17) did not have access to education. Meanwhile, when making a comparison between the years 2019 and 2020, the number of children out of school in South Africa increased by 81,961 (Statista, 2023). Thus, it is no surprise that Hensley (2017) and Stofleth (2016) indicated that achieving global sustainable development (SD) and Sustainable Development Goal (SDG) 4 may be problematic, especially in developing countries such as Brazil and South Africa.

In response to the above-mentioned imperatives, Brazil and South Africa have committed to the United Nations (UN) 2030 Agenda for Sustainable Development. Agenda 2030 views eradicating poverty, improving well-being, protecting human rights, and restoring the earth's balance as interconnected challenges (Statistics South Africa, 2019; World Economic Forum, 2015). To create a world that is sustainable and to provide equal opportunities for all, citizens need to be equipped with the necessary knowledge, skills, and attitudes, which can be accomplished through education, which is considered the most influential tool to achieve sustainability (UNESCO, 2013: 7). Moreover, Education for Sustainable Development (ESD) is essential to transform the value systems of citizens and to reshape the way citizens think about and act on economic, societal, and environmental challenges (UNESCO, 2014: 8).

Education for sustainable development has been reinforced by the UN SDG 4, as it focuses on inclusive, equitable, quality, lifelong education (Farinha et al., 2018: 915). Rampasso et al. (2020: 1) noted that ESD is linked to quality education (SDG 4), as stated in target 4.7; thus, education should be utilised to develop knowledge and values that enable action towards SD. Ferguson and Roofe (2020: 962) reported that target 4.3 pays special attention to higher education institutions (HEIs), as this target's aim is to have equal access to HEIs (technical, vocational, and tertiary education) for men and women by 2030. In addition, the realisation of SDG 4 is critical, as it has the potential to contribute to the achievement of the remaining 16 SDGs (Ferguson & Roofe, 2020: 960; Rampasso et al., 2020: 1).

Yet, it is concerning that the realisation of the SDGs in many HEIs is in the initial stage. UNICEF (2021) reported that the attainment of SDG 4 is imperative for two reasons. Firstly, SDG 4 encourages lifelong education and supports equality in terms of educational access and quality. Secondly, the long-term goal of sustainability relies on all children receiving quality education. When a child's potential is nurtured, they evolve into productive citizens, equipped to positively contribute to society. Ferguson and Roofe (2020) noted that the SDGs are interlinked; thus, one goal cannot be achieved without the other. Similarly, one level of education affects the other. HEIs need to share their power and prestige with other stakeholders at lower levels to have the most coordinated efforts to achieve SDG 4 at all levels (Ferguson & Roofe, 2020: 965).

Zooming in on the two contexts under investigation, legislation protecting and promoting SD in Brazil is plentiful; however, the most significant piece of legislation protecting the environment, human rights, and equality is the Constitution of the Federative Republic of Brazil (Brazil, 1998: 150). It is worth noting that Brazil played a leading role in the implementation of the Millenium Development Goals (MDGs). Consequently, Brazil attained a seat on the UN High-Level Group that monitors the implementation of Agenda 2030 (Brazilian Government, 2017: 9). In addition, Marques et al. (2018: 42) commended the proactive steps that academics in Brazil have taken to insert ESD into research and extension activities of HEIs. As is evident in the previous paragraph, scholars point to Brazil's commitment to sustainability. Scholarly literature, however, also identifies several barriers that hinder the progress made towards SD (Brandli et al., 2015; Cernov & Pietricovsky, 2020; Fearnside, 2018; Marques et al., 2018). For instance, while former president Jair Bolsonaro was in power, the Brazilian government had a negative impact on the protection of human rights and SD (Cernov & Pietricovsky, 2020; Fearnside, 2018). Cernov and Pietricovsky (2020) noted that Brazil has abandoned its commitment to most of the SDGs. Meanwhile, Brandli et al. (2015: 68) and Marques et al. (2018: 43) voiced their concern about the insertion of ESD, which has been slow in Brazil due to a lack of policies and motivation from HEIs' staff and upper management. On the same note, Brandli et al. (2015: 66) stated that action towards sustainability in Brazilian HEIs has been isolated and has not been contained in an organised vision or policy. Although Brazilian HEIs focused on sustainable actions, little attention was paid to ESD, as there was a lack of funding directed towards sustainable education (Brandli et al., 2015: 68).

An observation about the Brazilian context is that HEIs have a unique opportunity to develop problem-solvers and decision-makers that would balance social, economic, and environmental needs when making decisions at their places of employment (Rampasso et al., 2020: 2). Consequently, students need a profound understanding of the SDGs and the steps that should be taken to attain them (Ferguson & Roofe, 2020: 963). Furthermore, none of the SDGs can be achieved without HEIs being utilised as avenues through which citizens develop competencies to advance society (Ferguson & Roofe, 2020: 963; O'Malley, 2019).

Like Brazil, South Africa shows promising progress towards SD as well as pressing concerns that require attention. Scholarly literature (e.g., Constitution of the Republic of South Africa [RSA], 1996; Department of Environmental Affairs, 2019; Kimanzi, 2019; Letsema, 2019; Statistics South Africa, 2019; UNESCO, 2013) provides evidence that ESD in South Africa has grown in quality and scope, which is reflected in three areas, namely: (a) legislation and policies; (b) commitment to the SDGs; and (c) integration of ESD in HEIs.

Firstly, South African legislation and policies entail laws that reflect and promote key ESD issues such as human rights, environmental conservation, gender equality, peace building, and social justice (UNESCO, 2013: 56). One of the most significant pieces of legislation promoting SD is the Constitution of the Republic of South Africa (RSA, 1996: 6), which links environmental concerns with human rights and social responsibilities. Secondly, South Africa has participated in strategic planning meetings, programmes, and summits that aim to assist the country in achieving the 17 SDGs. To that end, in 2008, the South African Cabinet embraced the National Framework for SD, which aimed to declare South Africa's national vision to redirect development onto a sustainable path. In addition, South Africa participated in the Partnership for Action on Green Economy (PAGE) programme to position sustainability at the heart of economic policies and action (Department of Environmental Affairs, 2019). In 2015, South Africa linked the country's National Development Plan (NDP) with the 17 SDGs and 169 SD targets, resulting in 74% of the SD targets being addressed by the NDP (Statistics South Africa, 2019). Lastly, ESD has been incorporated into South African HEIs (UNESCO, 2013: 69). UNESCO (2013) commended Rhodes University and the University of Cape Town for their attempts to implement ESD in their curriculum, research, and extension activities. In addition, UNESCO (2013) applauded South African HEIs when a review of master's degrees, specifically in the field of environmental sciences and sustainability, revealed that 50% of HEIs in South Africa offered master's programmes in environment and sustainability sciences. Despite the promising progress made by South Africa towards ESD, there is still much to be accomplished. For instance, it is concerning that HEIs and businesses are poorly implementing legislation, policies, and strategies directed towards ESD (Letsema, 2019; UNESCO, 2013). In addition, undergraduate courses presented at South African HEIs do not pay enough attention to ESD (Kimanzi, 2019; UNESCO, 2013: 79). Further concerns highlighted by UNESCO (2013) are that although South Africa has a national development strategy towards SD (Statistics South Africa, 2019), to date, the country does not have a national ESD strategy and a coordinating body, which has disadvantaged the implementation of ESD. Additionally, there is a lack of long-term funding for ESD (UNESCO, 2013: 56).

Against this backdrop, based on systematic document analysis, this study brings attention to the path of BRICS (Brazil, Russia, India, China, South Africa) nations in the implementation of SDG 4. In addition, strategies to promote the realisation of target 4.7 and target 4.c, through ESD in the classroom are provided based on lessons learnt from Brazilian and South African HEIs. For the purpose of this study, the qualitative document analysis employed the principles of grounded theory and relied upon the purposeful selection of 60 documents from five Brazilian and five South African HEIs. The selected documents included strategic plans, mission and vision statements, institutional development plans, research strategies, and journal articles. In alignment with the ethical clearance granted for this study, the names of the institutions under investigation and thus the titles of the selected documents are kept anonymous. The two-fold research purpose was to (a) systematically examine relevant documents to explore the path trajectory of Brazil and South Africa in implementing SDG 4 and (b) provide recommendations for how HEIs in both contexts can enhance the implementation of ESD in the classroom. The following sections address the set objectives. It starts with a discussion of the path trajectory towards SDG 4 in Brazil and South Africa, charting the key challenges in implementing ESD towards the realisation of SDG 4. It then discusses the strategies educators can employ to enhance the implementation of ESD in the classroom.

2 Brazil and South Africa's Path Trajectory Towards SDG 4

Recognising the important role HEIs play in the attainment of SDG 4 and the remaining SDGs, this subsection explores the path trajectory of South Africa and Brazil towards the realisation of SDG 4 and its targets.

Special attention is paid to target 4.3, which aims to create equal access to HEIs for all citizens.

Brazil is faced with three challenges hindering progress towards the attainment of SDG 4 and its targets. Firstly, budget cuts have made free education for all Brazilian citizens unlikely (target 4.1). Secondly, legislation such as the National Education Plan is poorly implemented, and new legislation such as the Supplementary Law Bill 39/2020 does not promote SDG 4, as it makes provision for the appointment of new teachers (target 4.c) only if they do not increase expenses (Civil Society Working Group for the 2030 Agenda, 2020: 16). Lastly, there is a lack of reporting, statistics, data, and current research evaluating Brazil's progress towards SDG 4 (Civil Society Working Group for the 2030 Agenda, 2020: 16).

Progress towards free primary and secondary education (target 4.1) is insufficient, as 90,000 children between the ages of six and 14 do not attend school (Civil Society Working Group for the 2030 Agenda, 2020: 16). Furthermore, enrolment of children under the age of three in pre-primary education has improved from 10 to 23% between the years 2012 and 2017. However, this percentage (23%) remains below the OECD average of 36% (OECD, 2019: 1). As reported by the Civil Society Working Group for the 2030 Agenda (2020: 16), it is concerning that there is a lack of recent data, and no plans are in place to collect data on early childhood development enrolment rates, which have implications for equal access to quality pre-primary education (target 4.2).

Concerns extend from primary education to tertiary education, as only 15% of Brazilian citizens are enrolled in HEIs. Although post-schooling enrolment rates are low, on average, the Brazilian government has spent 10 times more of its financial resources on tertiary education than on primary education (OECD, 2019: 2; Pollack, 2017). Post-school education and training experienced steep budget cuts between the years 2013 and 2019 (1.17 billion to 2.46 million). This has reduced the enrolment rate in post-school education and training and has negatively affected the progress towards creating equal access to HEIs (target 4.3). According to the Civil Society Working Group for the 2030 Agenda (2020: 16), poor enrolment and completion rates have reduced the potential to equip citizens with skills for financial success (target 4.4). Furthermore, Pollack (2017) noted that while tuition-free public HEIs are funded by the taxes of the poor working class, citizens from affluent backgrounds do better in entrance exams, thus gaining access to HEIs (target 4.3, 4.5). In addition, the OECD (2019) reported that admission to vocational programmes is selective, and students from affluent backgrounds are more likely to be enrolled in vocational programmes than disadvantaged citizens (target 4.3, 4.5).

The Civil Society Working Group for the 2030 Agenda (2020) informs that it is difficult to report on target 4.5, as there is a lack of recent data on and research into the elimination of discrimination in education; however, historical inequalities still manifest within the Brazilian education system and society (target 4.5). Furthermore, achieving universal literacy and numeracy levels (target 4.6) by 2030 seems unlikely, as 11 million Brazilians were illiterate in 2019. In addition, the Civil Society Working Group for the 2030 Agenda (2020) notes that environmental education is no longer compulsory at all levels of education, which reveals Brazil's poor commitment to ESD and global citizenship (target 4.7).

Like Brazil, South Africa must overcome historical inequalities and cope with budget cuts to ensure SDG 4 and its targets are achieved. According to the Business Tech website (2021), the budget allocated to basic education is under pressure for two reasons. Firstly, the annual budget allocated towards education decreased by 5.3% between the years 2020 and 2021. Secondly, many parents of children attending fee-paying schools do not pay school fees. Meanwhile, the Business Tech website (2021) and Griffiths (2017) reported that two-thirds of the country's children attend no-fee primary schools and do not pay school fees, but those attending special schools often do (target 4.1, 4.5). In addition, the South African Government (2019) is aware that South Africa has one of the most unequal education systems in the world as a direct result of the apartheid system (target 4.5).

It is promising that quality education has been prioritised by South African legislation, such as the National Development Plan (NDP) (South African Government, 2019: 58). As is evident, 95% of the country's children are engaged in formal learning in the year before officially starting primary school (target 4.2). Moreover, 95% of South African citizens aged between 16 and 18 have passed Grade 7. Yet the dropout rates in senior grades and HEIs are high (South African Government, 2019: 59).

We learn from Matsolo et al. (2018: 63) that since 1994, South Africa has made tremendous headway towards improving enrolment rates in HEIs (target 4.3). Since the year 2000, enrolment rates have been increasing at an average rate of 4.2% annually. In addition, the South African government allocates a large portion (20%) of its annual budget to basic and tertiary education (Matsolo et al., 2018: 63; UNICEF, 2018: 4). Despite the government's commitment, South African HEIs have 1 901 tertiary education enrolments per 100 000 of the population (Nksoi, 2019). This is the lowest as part of the BRICS member states, with China at 3 104 and Brazil at 4 023 per 100 000 of their population (Nksoi, 2019; South African Government, 2019: 62). In addition, South African youth (aged between 18 and 24) continue to struggle to gain admission to the post-school education and training sector (South African Government, 2019: 62). In the year 2016, 2.9 million South African youth were enrolled at HEIs (Nksoi, 2019; South African Government, 2019: 62). However, 3.2 million citizens in the same age group were unemployed and had not engaged in any post-school education and training, which negatively affects the number of citizens capable of financial success (target 4.4).

In terms of teaching and learning conditions, all schools in South Africa have access to drinkable water and electricity, and sustained action is being taken to improve learning and teaching conditions. However, the South African Government (2019: 61) is currently concerned that South African rural schools offer teaching and learning experiences and resources that are below par (target 4.5). Promising progress has been made towards target 4.6, as almost 100% literacy levels have been reached by South African citizens aged between the ages of 15 and 64 (South African Government, 2019: 59). Use of the Internet and computers for the purpose of teaching and learning in South African schools is low compared to the international norm. As reported by the South African Government (2019: 59), South Africa could seek to partner with other countries well vested in the use of technology in the classroom to develop globally competent citizens (target 4.7).

Overall, it is promising that South Africa has identified the challenges the country needs to overcome and that the country has provided possible solutions. The country intends to achieve SDG 4 and its targets by 2030 through offering diverse training opportunities to educators (target 4.c) and other professions, providing entrepreneurial skills training, monitoring performance, and making necessary changes in order to address the socio-economic factors hindering enrolment and completion rates in the post-school education and training sector (South African Government, 2019: 62).

3 Challenges in Creating Effective Classroom Learning Environments Towards ESD

SDG 4 is directed at different levels of education, compromising of ten targets, seven of the targets make up the expected outcomes and three targets are strategies of achieving these targets (UNESCO, 2021). One of the targets (of SDG 4) is to ensure that through education, all learners acquire the competencies to take action towards SD (target 4.7). Meanwhile, one of the strategies to achieve SDG 4 is to create effective learning environments (target 4.a). This section will list the response of the two BRICS nations to challenges in creating effective classroom learning environments (target 4.a) when teaching towards ESD (target 4.7). Subsequently, the following section makes recommendations towards enhancing the implementation of ESD (target 4.7) by maximising effective classroom learning (target 4.a) through sound pedagogical approaches (target 4.c).

As per the challenges revealed by the document analysis, many academics do not see the link between ESD and pedagogical innovation, some academics view sustainability as being imposed on them and not linked to their respective disciplines, and an overcrowded and inflexible curriculum prevents academic staff from incorporating ESD into their teaching. However, this section reports on the following three main challenges from the two BRICS nations under investigation: A lack of understanding of the holistic nature of sustainability and related terms, inadequate pedagogical approaches that does not develop habits and mindsets that enable citizens to live sustainably and poorly specified curriculum content that does not capture the holistic nature of sustainability.

The first challenge is that educators and students do not fully understand the holistic nature of sustainability and related terms, therefore, it is poorly implemented in their teaching and other core activities. The term sustainability gives rise to numerous discussions and debates in a political and epistemological sense. The interdisciplinary nature of the concept and the lack of an agreed-upon definition hinder engagement in ESD in the Brazilian and South African contexts. In fact, the majority of the Brazilian and South African HEIs under investigation did not define what sustainability means in the context of their institution. In addition, the role of stakeholders (staff and students) towards sustainable development was not articulated. To this end, as in both BRICS contexts, a poor understanding of ESD's holistic and interdisciplinary nature led to its poor implementation in practice.

The second challenge is that pedagogical approaches are inadequate and do not develop habits and mindsets that enable citizens to live sustainably. ESD requires lifelong learning, the nurturing of globally competent citizens, and the use of transformative teaching pedagogies. Jodoin (2020: 779) expresses his concern that teaching approaches with the potential to develop sustainability-related competencies are not being utilised in the classroom because they conflict with traditional teaching approaches.

Zooming in on the two BRICS nations, the analysed documents reveal that the majority of the sampled HEIs in Brazil are prioritising transformative teaching methods and lifelong learning. In the case of one Brazilian HEI, sustainability-literate graduates are developed by nurturing globally competent citizens and employing student-centred approaches infused with real-life, sustainability-related critical thinking and problem-solving. Meanwhile, the rise of technology for educational purposes has enabled South African HEIs to adapt to the dynamic nature of sustainability. One South African HEI has adapted to the changing context of education in the twenty-first century, with a greater focus on the development of transferrable skills such as critical and creative thinking as opposed to memorising information. Furthermore, lifelong learning, the nurturing of globally competent citizens, the use of transformative teaching pedagogies, real-life, sustainability-related critical thinking, and problem-solving are employed in the classrooms of the majority of South African HEIs.

The fact that curriculum content is poorly specified and curriculum flexibility is limited leads to the curriculum not capturing the holistic nature of ESD, which accounts for the third challenge. The current reality of teaching is that single lectures accommodate many students—the curriculum is narrow, and assessment-focused. Consequently, usually, there is no room in the curriculum to introduce new SD content, as sustainability is not part of the official portfolio of courses. A curriculum that is receptive and adaptive to the dynamic nature of sustainability is needed for the successful implementation of ESD, which is evident in the case of Brazil. Curricula flexibility is understood and practised in order to respond to the challenges and ever-changing nature of society and sustainability. One Brazilian HEI is adapting to the dynamic nature of sustainability, as is evident by the modern curriculum being implemented. The curriculum is developed with the intent to personalise students' education and provide students with opportunities for a humanistic, democratic, interdisciplinary, and transdisciplinary education that values differences and cultural diversity. South African HEIs update the content in the curriculum and invest in teaching materials in order to adapt to the dynamic nature of sustainability. One South African HEI is of the belief that rigorous quality assurance of all academic programmes and courses, and continuously renewing curriculum content, is necessary to achieve their mission. Although it is clear that Brazilian and South African HEIs incorporate sustainability-related content in their curriculums, ESD is poorly specified and does not contain enough interdisciplinary and holistic detail to change the mentality of citizens.

4 Enhancing the Creation of Effective Learning Environments Towards ESD

Although capitalising on opportunities to maximise the implementation of ESD in the HEI as a whole is imperative (see Daniels & Niemczyk, 2023), attention also needs to be devoted to the classroom. It is recommended that academics, educators, and management of HEIs consider the following strategies to maximise the implementation of ESD in the classroom context. Based on the sampled documents, four strategies towards maximising the implementation of ESD in the classroom emerged:

i. Shift pedagogical approaches from teacher-centred methods and theoretical learning to student-centred approaches incorporating critical thinking and problem-solving;
ii. Include inter-, multi-, and transdisciplinary approaches, perspectives, and knowledge in the learning process;
iii. Cultivate sustainability-related knowledge, competencies, and values;
iv. Foster a democratic learning process.

The most important prerequisite to maximising ESD in the classroom is the pedagogical competencies of educators. Although a necessary prerequisite to developing sustainability-related values and competencies, classroom learning should not be limited to merely conveying expert sustainability knowledge to students. Instead, it should be coupled with the acquisition of knowledge, skill values, and competencies to respond to the complexities and uncertainties of the future. Considering the complex and holistic nature of SD, multiple pedagogical approaches are required, since single pedagogical approaches are often ineffective. The effective integration of ESD into the classrooms of HEIs calls for a shift in traditional pedagogical approaches to innovative pedagogical methods that link theory and practice. More specifically, this would require a move away from teacher-centred approaches and theoretical learning towards (a) student-centred approaches, (b) critical thinking, and (c) problem-solving.

The employment of student-centred approaches will empower students to be co-constructors in the learning process. Student-centred approaches will require students and educators to commit to collaboration, openness, trust, and patience, which will take time. Student-centred approaches not only encourage participation; they also create space for deep transformative learning, which, in turn, creates a favourable platform for students to develop reflective skills, to enquire and break away from the status quo, as well as to question existing unsustainable practices and values. Student-centred pedagogical approaches to SD should be coupled with critical reflection, action-oriented learning, critical reflective thinking, participatory approaches, inter- and transdisciplinary approaches, and systemic thinking. To ensure optimal effectiveness, changes at two interconnected levels of pedagogy should occur. The first change is in the classroom, which will require lecturers to incorporate transformative educational practices such as critical thinking and problem-solving. The second change is embedded in the curriculum, as it should be flexible and orientated towards ESD. Zooming in on the criterion for lecturers to evoke critical thinking is based on the reality that critical thinking has the potential to nurture a critical attitude towards the values of ESD. In addition, instilling in students the competency to think critically will promote confidence when confronting challenges that hinder transformation and sustainability. To this end, transformation and sustainability should be steered by pedagogical approaches that enable students to transform their current realities through critical thinking and real-life problem-solving. The incorporation of real-life problem-solving into teaching will require the simulation of real-life, local, and global sustainability-related problems. One notable method of simulating real-life sustainability scenarios in teaching is to expand the setting of teaching beyond the classroom. Outdoor learning serves as an effective setting for the acquisition of sustainability-related knowledge, skills, and actions. Outdoor learning can occur in the form of extracurricular activities, excursions, and service learning. The knowledge, values, and competencies attained through the outdoor experiences should be linked with classroom theory in order to achieve praxis. Extracurricular activities and service learning as transformative pedagogical approaches provide students with a space (a) to learn and understand the theory in which ESD is embedded and (b) to engage in actions that benefit society and the environment.

Extracurricular activities and service-learning expose students to a variety of sectors within society, facilitating inter-, multi-, and transdisciplinary approaches to teaching and learning. These approaches are necessary due to the complex, interconnected nature of sustainability, requiring complex competencies. It is also important to note that educators need to constantly cultivate sustainability knowledge, skills, and values, as SD is a lifelong learning process.

Furthermore, as regards the effective implementation of ESD in the classroom, the cultivation of sustainability-related (a) knowledge, (b) competencies, and (c) values is imperative. Sustainability-related knowledge should include a wide range of topics and perspectives connected in a multi-, inter-, and transdisciplinary manner. Topics should include, but not be limited to, economic development, educational improvement, equal health care, climate change, energy systems, biodiversity, social justice, and urbanisation. It is also important that students uncover the interconnections between the different dimensions of sustainability (environmental, development, social, economic, cultural) and the interconnectedness of sustainability-related knowledge (multi-, inter-, and transdisciplinary). Uncovering the holistic and complex nature of sustainability and the different sectors of society enhances the capacity to think critically. One notable method of incorporating expert sustainability-related knowledge into courses is to capitalise on technological advancements. Courses can rely on data, videos, and live seminars with global experts to acquire sustainability-related knowledge and become informed citizens.

As stated earlier, knowledge about SD is not enough; students also need to develop appropriate competencies and values. It is important to note that prescribing a mandatory set of competencies required for SD is beyond the scope of this study. This is due to the reality that there are multiple definitions and perspectives about sustainability and competence in an educational context. The following ESD competencies were identified in the analysed documents:

- Considering changes in the future, and thinking about potential alternative scenarios in order to adapt;
- Putting challenges and opportunities into context, taking into consideration the different dimensions, which include the physical dimension (local–global) and the temporal dimension (past, present, and future);
- Innovating in times of uncertainty by using multi-, inter- and transdisciplinary strategies;
- Making decisions using critical thinking and problem-solving to question assumptions while maintaining an open mind and respect for different types of perspectives and cultures;
- Participate, plan, and implement strategies and decisions;
- Possessing the emotional intelligence to reflect, manage emotions, and feel empathy, sympathy and solidarity;
- Considering different knowledge sources and perspectives before making conclusions, decisions, and acting towards change.

At the heart of ESD are values because they serve as a driving force that will motivate citizens to lead more sustainable lives. Values that should be promoted by ESD are respect, commitment, and empathy in the context of the environment, heritage, and society at large. It is important to note that a focus on values in the educational process does not suggest an uncritical acceptance of sustainability values. Instead, the adoption of values should be examined, debated, and tested, which would lead to a process of critical engagement with key sustainability-related values. Engaging students, educators, and stakeholders to challenge the values that are in line and in contrast with the ESD agenda creates a platform for the transformation of values-on-paper into values-in-action. This transformation will require students to internalise the values deemed necessary for transformation and sustainability. In turn, this could become a powerful driving force in the transformation of society towards sustainability. However, nurturing values-in-action through changing attitudes, values and behaviours is a challenging task and one that requires more than just knowledge dissemination. It is important to note that the internalisation of ESD values may not always transmit to transformation or sustainability. Optimal potential for transformation and sustainability occurs when critical thinking, problem-solving, and values are nurtured in an integrated fashion. An innovative teaching approach to instilling ethics and values is to provide students with the opportunity to partake in service learning, which enables them to learn and develop values through their active engagement in the local community. In addition to nurturing values required for SD, service learning has the potential to cultivate compassion, social skills, teamwork, and emotional intelligence.

The social principles of sustainability are grounded in democratic practices of equity, participation, and collective decision-making. In addition, education plays a pivotal role in the transformation and sustainability of society. For these two reasons, ESD needs to be democratic in nature. Democratic education views the engagement and participation of all stakeholders on an equal footing; thus, it upholds the value and right of equality for all. The contribution of democratic education towards ESD is embedded in the platform it creates for the involvement of its stakeholders (staff, students, and managers) in collective decision-making about social transformation and sustainability. Furthermore, the analysed documents identified four reasons as to why democratic education is aligned with ESD. Firstly, it assures social cohesion through sound ESD practices and community-based decision-making. Secondly, it assures the legitimacy of decisions and warrants that the agenda of transformation and sustainability are aligned with the needs of the community. Thirdly, democratic education gives a voice to previously silenced, marginalised and alternative voices. Lastly, it creates the space in which to challenge the unsustainable practices that could potentially maintain and advance unsustainability.

The development of sustainability-related competencies is especially important for educators to have a sustainability mindset and to carry cognisance of sustainability-related principles. Consequently, HEIs should prioritise programmes that develop sustainability-related knowledge, competencies and values in Faculties of Education and beyond. Although Faculties of Education are important, ESD needs to be holistic and applied in all faculties. Thus, HEIs should continuously provide their teaching staff and support staff with ongoing professional support in developing their own ESD competence. In turn, this could transform teaching pedagogies that develop graduates with the potential to initiate transformation and take action towards a more just, sustainable society.

5 Conclusion

BRICS nations have shown considerable commitment and progress towards the UN 2030 Agenda for Sustainable Development and ESD through international dialogue, technology, conferences, teaching and learning, research, strategic planning, and signing of declarations (Farinha et al., 2020: 467; Mula et al., 2017: 799). However, the reality in many developing countries, such as South Africa and Brazil, is that ESD is only starting to gain academic momentum (Solera & Laya, 2017: 279). An obvious starting point towards improvement is to ensure students, staff, and stakeholders have a clear understanding of what SD means and their part in working towards it (Ferguson & Roofe, 2020: 964; Da Silva et al., 2019: 238; Tierney et al., 2016: 510).

As per the success and challenges towards SDG 4 and its targets in the BRICS nations, despite budget cuts, Brazil and South Africa have improved their enrolment rates. Although the literacy levels are high in South Africa, large portions of the working force (especially the youth) are unemployed. If SDG 4 is to be realised by 2030, Brazilian and South African governments and educational leaders should further improve enrolment rates (at all levels of education), address budget cuts, and tackle inequalities and disparities that continue to manifest in the respective education systems.

In terms of classroom practices geared towards target 4.7 and target 4.c, South African and Brazilian educators employed in HEIs grapple to nurture citizens to take action towards SD. The main classroom challenges are a lack of understanding of the holistic nature of sustainability and related terms, inadequate pedagogical approaches that do not develop habits and mindsets that enable citizens to live sustainably, and poorly specified curriculum content that does not capture the holistic nature of sustainability. As is evident in the reviewed literature, HEIs can respond to these challenges and increase their efforts towards the attainment of SDG 4 and its targets by shifting pedagogical approaches from teacher-centred methods and theoretical learning to student-centred approaches and critical thinking; incorporating critical thinking and problem-solving, include inter-, multi- and transdisciplinary approaches, perspectives and knowledge in the learning process; cultivating sustainability-related knowledge, competencies, and values and fostering a democratic learning process. Developing sustainability-related competencies of educators, educational leaders, and students through professional development and teacher training courses holds high potential to effectively respond to several challenges and to shift the world onto a sustainable path.

The limitations of the findings in this study include the reliance on qualitative document analysis as the primary research method, which may have limited the depth of insight that could have been gained from direct interactions with stakeholders such as educators, students, and policymakers. Future studies could address this limitation by incorporating interviews, surveys, or focus groups to gather perspectives and experiences directly from individuals involved in the implementation of ESD. Additionally, expanding the scope of the study to include a larger sample size of HEIs from a more diverse range of countries beyond Brazil and South Africa could provide a broader understanding of the challenges and strategies related to the realisation of SDG 4 and its targets.

References

Brandli L. L., Leal Filho, W., Frandoloso, M. A. L., Korf, E. P., & Daris, D. (2015). The environmental sustainability of Brazilian universities: Barriers and pre-conditions. In W. Leal Filho, U. Azeiteiro, S. Caeiro, & F. Alves (Eds.), *Integrating sustainability thinking in science and engineering curricula* (pp. 63–74). World Sustainability Series. Springer. https://doi.org/10.1007/978-3-319-09474-8_5

Brazil. (1998). *Constitution of the Federative Republic of Brazil of 1998.*

Brazilian Government. (2017). *Brazil debates sustainable development goals.* Retrieved January 8, 2021, from http://www.brazil.gov.br/about-brazil/news/2017/04/brazil-debates-sustainable-development-goals

Business Tech. (2021). *A growing number of schools in South Africa have a real problem—Unpaid school fees.* Retrieved May 8, 2021, from https://businesstech.co.za/news/finance/463926/a-growing-number-of-schools-in-south-africa-have-a-real-problem-unpaid-fees/

Cernov, A., & Pietricovsky, L. (2020). *Brazil at the crossroads: The SDGs, the crises and the impact of Covid-19.* Retrieved January 7, 2021, from https://www.socialwatch.org/node/18506

Civil Society Working Group for the 2030 Agenda. (2020). *2030 agenda for sustainable development spotlight report synthesis IV Brazil.* Retrieved January 24, 2021, from https://brasilnaagenda2030.files.wordpress.com/2020/07/en_osd_2020_vs_2_mari_singlepage.pdf

Da Silva, R., Tommasetti, R., Gomes, M., & Macedo, M. (2019). How green is accounting? Brazilian students' perception. *International Journal of Sustainability in Higher Education, 21*(2), 228–243. https://doi.org/10.1108/IJSHE-07-2019-0232

Daniels, C., & Niemczyk, E. (2023). Towards higher education for sustainable development in BRICS: Focus on Brazil and South Africa. *Space and Culture, India, 11*(1), 68 77. https://doi.org/10.20896/saci.v11i1.1354

Department of Environmental Affairs. (2019). *Advancing inclusive and sustainable economies: A report on the 2019 PAGE ministerial conference.* Retrieved

January 9, 2021, from https://www.environment.gov.za/sites/default/files/reports/page-conference-report2019.pdf

Farinha, C., & Caeiro, S. S. (2018). Education for sustainable development in Portuguese universities. *International Journal of Sustainability in Higher Education, 19*(5), 912–941. https://doi.org/10.1108/IJSHE-09-2017-0168

Farinha, C., Caeiro, S., & Azeiteiro, U. (2020). Universities speak up regarding the implementation of sustainable development challenges: The case of Portugal. *International Journal of Sustainability in Higher Education, 21*(3), 465–506. https://doi.org/10.1108/IJSHE-07-2019-0232

Fearnside, P. (2018). *Yale environment: Why Brazil's president poses an unprecedented threat to the Amazon.* Retrieved January 12, 2021, from https://e360.yale.edu/features/why-brazils-new-president-poses-an-unprecedented-threat-to-the-amazon

Ferguson, T., & Roofe, C. (2020). SDG 4 in higher education: Challenges and opportunities. *International Journal of Sustainability in Higher Education, 21*(5), 959–975. https://doi.org/10.1108/IJSHE-12-2019-0353

Friedman, S. (2020). *South Africa is struggling with Covid-19 because their leaders want to emulate the first world.* Retrieved January 29, 2021, from https://qz.com/africa/1882297/south-africa-struggles-with-covid-19-as-it-shows-inequalities/

Griffiths, H. (2017). *Free primary education still out of reach in South Africa.* Retrieved May 8, 2021, from https://www.hrw.org/news/2017/04/13/free-primary-education-still-out-reach-south-africa

Hensley, N. (2017). *The future of sustainability in Higher Education.* Retrieved January 7, 2021, from http://www.susted.com/wordpress/wp-content/uploads/2017/03/Hensley-JSE-March-2017-Future-Casting-Issue-PDF1.pdf

Human Rights Watch. (2021).*Brazil: Failure to respond to education emergency.* Retrieved May 8, 2021, from https://www.hrw.org/news/2021/06/11/brazil-failure-respond-education-emergency

Jodoin, J. J. (2020). Promoting language education for sustainable development: A program effects case study in Japanese higher education. *International Journal for Sustainability in Higher Education, 21*(4), 779–798. https://doi.org/10.1108/IJSHE-09-2019-0258

Kimanzi, M. K. (2019). Education for Sustainable Development: Economics students perspectives at an institution of higher learning in South Africa. *International Journal of eBusiness and eGovernment Studies, 11*(1), 53–67. https://doi.org/10.3390/su11216136

Letsema. (2019). *Four reasons why ESD programmes fail.* https://letsema.co.za/consulting/economic-development/enterprise-development/four-reasons-esd-fail/. Date of access: 12 January 2021.

Marques, C. S., Trindade, N. R., Favarin, R. R., Telocken, S. G., & Trevisan, M. (2018). Reflections on sustainable practices: Analysing teaching, research, extension and management of a Brazilian Public University. In W. Leal Filho, F. Frankenberger, P. Iglecias, & R. Mülfarth (Eds.), *Towards green campus operations: World sustainability series*. Springer. https://doi.org/10.1007/978-3-319-76885-4_2

Matsolo, M., Ningpuanyeh, W., & Susuman, A. (2018). Factors affecting enrolment rate of students in higher education institutions in the Gauteng province, South Africa. *Journal of Asian and African Studies, 53*(1), 63–80.

Mula, I., Tilbury, D., Ryan, A., Mader, M., Dlouha, J., Mader, C., Benyas, J., Dlouhy, J., & Alba, D. (2017). Catalysing change in higher education for sustainable development: A review of professional development initiatives for university educators. *International Journal of Sustainability in Higher Education, 18*(5), 798–820. https://doi.org/10.1108/IJSHE-03-2017-0043

Nksoi, B. (2019). *Few SA youths go to varsity compared to those in BRICS countries.* Retrieved May 7, 2021, from https://www.iol.co.za/the-star/news/few-sa-youths-go-to-varsity-compared-to-those-in-brics-countries-21549701

OECD. (2019). *Education at a glance 2019.* Retrieved June 2, 2021, from https://www.oecd.org/education/education-at-a-glance/EAG2019_CN_BRA.pdf

O'Malley, B. (2019). *University World News: SDGs not attainable without contribution from HE.* Retrieved January 9, 2021, from https://www.universityworldnews.com/post.php?story=20190719135507840

Pollack, N. (2017). *Case study: Brazil's free education.* Retrieved May 8, 2021, from https://mg.co.za/article/2017-03-31-00-case-study-brazils-free-education/

Rampasso, I., Quelhas, O., Anholon, R., Pereira, M., Miranda, J., & Alvarenga, W. (2020). Engineering education for sustainable development: Evaluation criteria for Brazilian context. *Sustainability, 12*, 1–11. https://doi.org/10.3390/su12103947

Republic of South Africa. (1996). *The Constitution of the Republic of South Africa 108 of 1996.*

Solera, C. R., & Laya, M. (2017). Higher education for sustainable development at EARTH University. *International Journal of Sustainability in Higher Education, 18*(3), 278–293. https://doi.org/10.1108/IJSHE-06-2015-0104

South African Government. (2019). *South Africa's implementation of the 2030 agenda for sustainable development.* Retrieved May 5, 2021, from https://sustainabledevelopment.un.org/content/documents/23402SOUTH_AFRICA_RSA_Voluntary_National_Review_Report_Final__14_June_2019.pdf

Statista. (2023).*South Africa: children out of school 2001–2020*. Retrieved January 7, 2021, from https://www.statista.com/statistics/1261212/number-of-children-out-of-school-in-south-africa/

Statistics South Africa. (2019). *Sustainable development goals: Country report 2019*. Retrieved January 7, 2021, from http://www.statssa.gov.za/MDG/SDGs_Country_Report_2019_South_Africa.pdf

Stofleth, D. (2016). *Rethinking prosperity: A short history of sustainable development*. Retrieved January 10, 2021, from http://rethinkingprosperity.org/a-short-history-of-sustainable-development/

Tierney, A., Tweddell, H., & Willmore, C. (2016). Measuring education for sustainable development: Experiences from the University of Bristol. *International Journal of Sustainability in Higher Education, 17*(6), 853–874. https://doi.org/10.1108/IJSHE-07-2013-0083

UNESCO. (2013). *UNESCO: National Journeys towards Education for Sustainable Development*. Retrieved January 11, 2021, from https://sustainabledevelopment.un.org/content/documents/925unesco8.pdf

UNESCO. (2014). *Roadmap for Implementing the Global Action Programme on Education for Sustainable Development*. Retrieved January 10, 2021, from https://sustainabledevelopment.un.org/content/documents/1674unescoroadmap.pdf

UNESCO. (2021). *Sustainable development goal 4 and its targets*. Retrieved July 15, 2023, from https://en.unesco.org/education2030-sdg4/targets

UNICEF. (2018). *Education budget brief*. Retrieved May 8, 2021, from https://www.unicef.org/esa/media/2536/file/UNICEF-South-Africa-2018-2019-Education-Budget-Brief.pdf

UNICEF. (2021).*SDG goal 4: Quality education*. 2021. Retrieved July 15, 2023, https://data.unicef.org/sdgs/goal-4-quality-education/

World Economic Forum. (2015). *Why education is the key to sustainable development*. Retrieved January 29, 2021, from https://www.weforum.org/agenda/2015/05/why-education-is-the-key-to-sustainable-development/

CHAPTER 5

Climate Change and Sustainable Development in BRICS Economies: A South African Policy Perspective

Basiiraa Moosa

1 Introduction

BRICS countries (Brazil, Russia, India, China, and South Africa) face shared challenges in addressing climate change and promoting sustainable development. As major economies and significant emitters, their actions have a significant global impact, while they collaborate on certain initiatives, their policy approaches and priorities differ due to varying economic and social conditions. Nevertheless, their commitment to emission reduction, renewable energy, and sustainable practices holds potential for influencing global efforts to combat climate change and foster sustainable development (Hau et al., 2012).

B. Moosa (✉)
Department of Finance and Investment Management, University of Johannesburg, Johannesburg, South Africa
e-mail: basiiraam@uj.ac.za

N. Tshikovhi et al. (eds.), *The BRICS Imperative*,
https://doi.org/10.1007/978-3-032-05932-1_5

Climate change and sustainable development are among the most critical global challenges of the twenty-first century, calling for immediate and comprehensive action from nations worldwide to act urgently on global challenges that demand immediate and comprehensive action from all nations (Ventura, 2023). The consequences of inaction are severe, affecting people, ecosystems, and economies worldwide. In the call to mitigate climate change transitioning to clean energy and adopting sustainable practices which influence sustainable development will result in economic growth, social progress, and environmental protection (Global Ocean Forum, 2010). Collaboration and solidarity are essential, and all stakeholders must work together to secure a sustainable and climate-resilient future for the planet (Myhre, 2013).

The BRICS countries, represent a significant portion of the world's population, landmass, and economic output. The BRICS countries collectively represent a significant portion of the world's population, landmass, and economic output. With over 40% of the world's total population, their large and rapidly growing populations have a profound impact on global demographic trends and social dynamics. The BRICS nations' combined land area covers approximately 26% of the Earth's land surface, encompassing diverse landscapes, climates, and ecosystems (Meena, 2013). This vast expanse makes them crucial players in terms of natural resources, environmental conservation, and biodiversity protection. Economically, the BRICS countries contribute significantly to the global economy, accounting for over 23% of the world's Gross Domestic Product (GDP) (International Monetary Fund, 2023) (Fig. 1).

As major emerging economies, they play a pivotal role in shaping international trade, investment, and economic growth with collective economic power giving them considerable influence in the geopolitical landscape and the formulation of global economic policies (International Monetary Fund, 2023). Given their substantial populations, extensive landmass, and significant economic clout, the BRICS countries' actions and policies carry weight in addressing global challenges (Santana et al., 2014). Issues such as climate change, sustainable development, poverty alleviation, and international trade are areas where the BRICS nations' collaboration and cooperation are crucial. Their concerted efforts have the potential to drive positive change on a global scale, impacting not only their respective regions but also influencing the course of the world's socio-economic and environmental future. As emerging economies, they

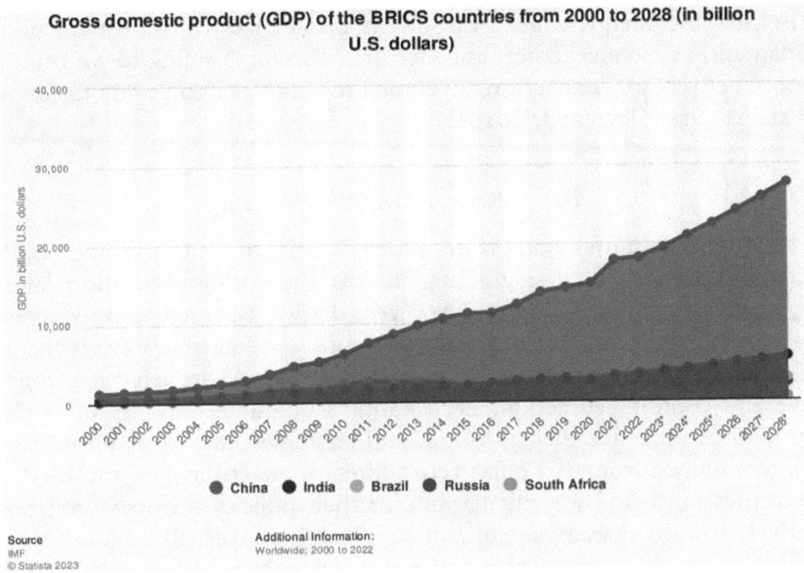

Fig. 1 GDP of the BRICS Countries from 2000 to 2028, IMF (2023)

play a crucial role in shaping the trajectory of climate change and sustainable development on a global scale (O'Neill, 2001).

South Africa has shown a strong commitment to addressing climate change and promoting sustainable development through its policy perspective. The country is particularly vulnerable to the impacts of climate change, such as droughts, heatwaves, and extreme weather events, which pose significant challenges to its economy, environment, and social well-being (UN Climate Summit, 2014). As a responsible global actor, South Africa acknowledges the importance of collective action and international cooperation in combating climate change (Ventura, 2023). South Africa's policy perspective on climate change and sustainable development is founded on the National Climate Change Response Policy (NCCRP) (Grantham Research Institute, 2011), emphasizing a transition to a low-carbon economy, renewable energy promotion, and adaptive capacities enhancement aligned with the SDGs, it integrates economic, social, and environmental dimensions through the National Development Plan (NDP). The Renewable Energy Independent Power Producer Procurement Programme (REIPPPP) drives private investment in renewable energy, while a carbon tax incentivizes emission reduction. Adaptation strategies target resilience in key sectors, aiming to contribute constructively to global efforts in combating climate change and fostering sustainability (Haywood, 2016).

1.1 *Recognizing the Problem*

The BRICS countries, comprising Brazil, Russia, India, China, and South Africa, collectively hold a significant share of the world's population, landmass, and economic output. However, despite their common interest in combating climate change and promoting sustainable development, divergent policy approaches have created challenges in achieving cohesive and coordinated action. Each nation's unique economic and social conditions influence their priorities and strategies, making it difficult to forge a unified front. To effectively address these challenges, the BRICS countries must find ways to harmonize their policies and work collaboratively toward shared climate and sustainability goals. By aligning their efforts, they can leverage their collective influence to drive global action against climate change and foster sustainable development practices. This coordinated approach holds the potential to create a more resilient and sustainable future for the BRICS nations and, in turn, contribute to global efforts in securing a healthier and more sustainable planet for all.

The potential collaboration within BRICS offers combined expertize, resources, and influences to address complex global challenges like climate change. It fosters learning, inclusivity, and resilience while optimizing resource efficiency. By working together, diverse stakeholders can achieve more sustainable and lasting impact, creating a better future for all.

This chapter aims to achieve three key objectives. Firstly, it seeks to analyze and compares the climate change and sustainable development policies of the BRICS countries to identify their similarities, differences, and potential areas of convergence. Secondly, it investigates the specific challenges and barriers that South Africa faces in its efforts to mitigate climate change and promote sustainable development within the BRICS framework. Lastly, the chapter aims to propose strategic approaches and policy recommendations that could enable South Africa to align more effectively with its BRICS counterparts, thereby fostering enhanced cooperation and coordination in addressing climate change and advancing shared sustainable development goals.

2 Main Body

The body of literature reveals that among the BRICS countries, there exists a mix of shared objectives and divergent approaches in their climate change and sustainable development policies. For instance, Brazil has been widely recognized for its commendable efforts in reducing deforestation rates and promoting the adoption of renewable energy sources. In contrast, Russia encounters distinct challenges in reducing emissions due to its heavy reliance on fossil fuels for its energy needs. India has made notable strides in the adoption of renewable energy technologies, while China's ambitious targets for emissions reduction demonstrate a strong commitment to climate action. South Africa, as a nation vulnerable to the impacts of climate change, has put in place a series of policies aimed at promoting renewable energy and bolstering adaptation strategies to mitigate the effects of environmental changes (Kıprızlı, 2023).

2.1 *Insights from Brazil*

Brazil has earned widespread recognition and acclaim for its notable achievements in the reduction of deforestation rates and its proactive efforts in promoting the adoption of renewable energy sources. As one of the BRICS countries, Brazil has been at the forefront of combating deforestation, particularly in the iconic Amazon rainforest. Through the implementation of robust environmental policies and conservation initiatives, the country has made significant strides in curbing deforestation activities, safeguarding invaluable biodiversity, and preserving one of the world's most critical ecosystems (Mahalik, Do educational levels influence the environmental quality? The role of renewable and non-renewable energy demand in selected BRICS countries with a new policy perspective., 2021).

Brazil has been highly proactive in embracing renewable energy solutions as part of its sustainable development agenda (Grantham Research Institute on Climate Change & the Environment, 2022). The country has strategically invested in and promoted the development of various renewable energy technologies, including wind, solar, and hydropower. These forward-thinking initiatives have not only helped to diversify Brazil's energy mix but have also substantially reduced the nation's reliance on fossil fuels. By leveraging the vast potential of clean and renewable energy sources, Brazil is making remarkable strides in mitigating its environmental impact while contributing to global efforts in combating climate change (Grantham Research Institute, 2011).

Through its commendable actions in both deforestation reduction and renewable energy promotion, Brazil serves as an exemplary model for other nations worldwide, showcasing the positive outcomes that can be achieved through a steadfast commitment to sustainable practices and responsible environmental stewardship. The country's achievements not only contribute to the well-being of its own population and ecosystems but also underscore its crucial role in advancing the global agenda for a greener, more sustainable future, by serving as a beacon of environmental responsibility and sustainable development within the BRICS group and beyond, Brazil continues to inspire and influence international efforts in addressing pressing climate and environmental challenges (UN Climate Summit, 2014).

2.2 *Insights from Russia*

Russia faces considerable challenges in reducing emissions primarily due to its heavy reliance on fossil fuels for meeting its energy needs. As one of the BRICS countries, Russia possesses vast reserves of oil, natural gas, and coal, which have historically been the backbone of its energy sector. However, this significant dependence on fossil fuels presents a complex conundrum for the country in its pursuit of emissions reduction and sustainable development goals. The fossil fuel industry has played a pivotal role in Russia's economic growth and global influence, contributing significantly to its GDP and export revenues. As a result, transitioning away from fossil fuels entails multifaceted economic, social, and political implications. The country must navigate a delicate balance between reducing emissions and maintaining energy security, economic stability, and energy affordability for its population. According to Climate Action Tracker (2023), the Russian Federation are showing insufficient action toward mitigating emissions (Fig. 2).

Despite the challenges, Russia has the stance that their National Development Plan influences the undertaking of efforts to address emissions reduction. The country is involved in various international agreements and initiatives aimed at mitigating climate change, such as the Paris Agreement. However, the pace of progress has been slower than desired due to

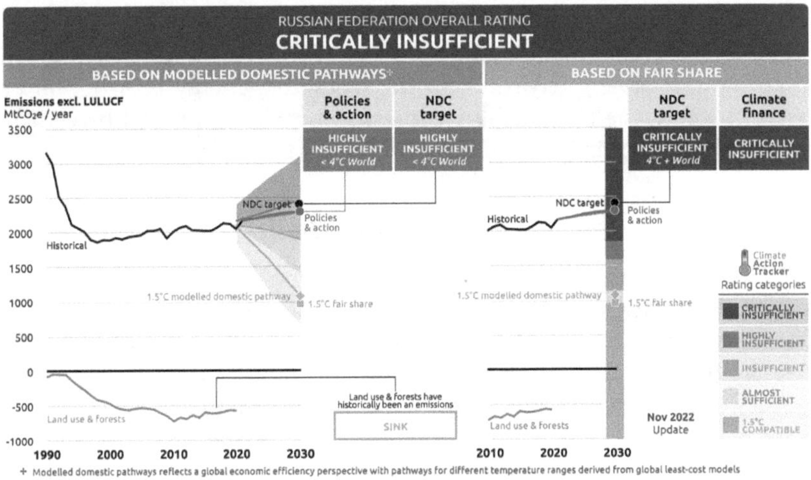

Fig. 2 Russian Federation overall rating, CAT (2023)

the complexity of reducing emissions while ensuring energy supply reliability and supporting traditional industries. The global energy landscape and geopolitical dynamics further influence Russia's approach to emissions reduction. As a major exporter of fossil fuels, Russia's position in the global energy market has both economic advantages and responsibilities (Lebedev, 2022). The country faces pressures and opportunities related to balancing domestic energy demand, export markets, and international commitments on reducing greenhouse gas emissions.

Considering these challenges, the Russian government has initiated steps to diversify its energy mix and embrace cleaner energy alternatives gradually through investments in nuclear power and renewable energy sources, such as wind and solar, some progress should be noted. However, scaling up these initiatives to significantly reduce emissions requires concerted efforts, innovative policies, and substantial investments in low-carbon technologies (Lebedev, 2022). Russia's path to emissions reduction represents a complex balancing act, wherein it seeks to harmonize environmental concerns with broader economic and energy security interests. As the global discourse on climate change intensifies, Russia's role in the transition to a more sustainable energy future will remain subject to scrutiny and calls for greater ambition in its emissions reduction commitments. Encouraging a shift away from fossil fuel reliance and fostering an enabling environment for renewable energy growth will be essential for Russia to effectively address the challenges posed by climate change while ensuring a sustainable and prosperous future.

2.3 *Insights from India*

India has emerged as a shining example in the global efforts to adopt renewable energy solutions and transition toward a sustainable and green future. As one of the BRICS countries, India has demonstrated remarkable progress and dedication in harnessing the power of renewable energy sources to meet its growing energy demands while mitigating the adverse impacts of climate change.

In recent years, India has experienced a significant increase in renewable energy capacity, particularly in the areas of solar and wind power. The country has actively invested in and promoted renewable energy projects, making it one of the world's leaders in solar energy deployment. Through ambitious targets and supportive policies, India has successfully attracted private investments, both domestically and internationally, in the renewable energy sector. The Government of India's commitment to renewable energy is evident in its ambitious initiatives, such as the Jawaharlal Nehru National Solar Mission (JNNSM), which aims to achieve 100 gigawatts (GW) of solar power capacity by 2022 (International Energy Agency, 2021). Additionally, the government has implemented various incentives and policy frameworks to encourage renewable energy adoption, including feed-in tariffs, tax incentives, and regulatory support for renewable energy projects.

India's impressive progress in adopting renewable energy has not only contributed to diversifying its energy mix but has also positively impacted its efforts in reducing greenhouse gas emissions. The increased use of renewable energy sources has resulted in a significant reduction in the country's carbon footprint, aligning with its commitments under the Paris Agreement (UN-Habitat, 2023) and demonstrating India's dedication to combating climate change. The adoption of renewable energy has not been limited to large-scale projects. India has also made substantial strides in promoting decentralized renewable energy solutions, such as solar rooftop installations and community-led initiatives. These efforts not only increase energy access for remote and underserved communities but also empower them to become active participants in the sustainable energy transition. India's renewable energy success story is supported by the growth of the clean energy industry, leading to job creation and economic opportunities. The renewable energy sector has become a key driver of employment, providing numerous jobs in manufacturing, installation, maintenance, and research and development (India Meteorological Department, 2023).

India's commitment to renewable energy is not only confined to its borders but also extends to international cooperation. The country actively engages in global partnerships, initiatives, and forums to share knowledge, best practices, and technology in the field of renewable energy. This collaboration fosters a spirit of collective action and supports the global agenda for sustainable development and combating climate change (Transforming our world: the 2030 Agenda for Sustainable Development, 2023). India's exemplary progress in adopting renewable energy demonstrates its dedication to a greener and more sustainable future. As a leading global player in the clean energy transition, India continues to inspire and influence other nations to prioritize renewable energy solutions and embrace sustainable practices. By leveraging its renewable energy potential, India showcases the transformative power of clean energy in building a resilient, low-carbon, and prosperous future for its citizens and the world at large.

2.4 *Insights from China*

China, as a pivotal member of the BRICS group, has taken on a leadership role in the global fight against climate change by committing to ambitious emissions reduction targets. The country's steadfast dedication to addressing its environmental challenges is evident through its comprehensive policy initiatives and proactive measures aimed at curbing greenhouse gas emissions and fostering sustainable development. Over the years, China has experienced significant economic growth, leading to an increased demand for energy and a corresponding rise in carbon emissions, however, recognizing the urgent need to address climate change, China has embarked on an unprecedented journey to transition toward a more sustainable and low-carbon future (Crompton & Wu, 2005). One of the key milestones in China's commitment to emissions reduction is its nationally determined contribution (NDC) under the Paris Agreement. The country pledged to peak its carbon dioxide emissions by 2030, and importantly, to achieve carbon neutrality by 2060. This ambitious goal signals China's strong determination to take concrete and transformative actions to mitigate its impact on the global climate.

To achieve these targets, China has embraced a multifaceted approach that includes comprehensive policy frameworks, extensive investment in clean energy technologies, and the implementation of ambitious renewable energy projects. The Chinese government has actively promoted renewable energy sources such as solar, wind, hydro, and nuclear power, resulting in significant capacity additions in these sectors (IEA, 2012). Additionally, China leads the world in solar and wind power installations, further solidifying its position as a global renewable energy powerhouse. The country's commitment to ambitious emissions reduction is further exemplified by its aggressive phasing out of coal-fired power plants and increased focus on energy efficiency measures. China has set stringent emissions standards for industries, enforced stricter pollution controls, and promoted the adoption of cleaner technologies across various sectors. These measures have had a notable impact in reducing the carbon intensity of China's economy (Zhou, 2011).

China's dedication to tackling climate change extends beyond its borders. The country has actively engaged in international collaborations and partnerships to share its experiences, technology, and best practices with other nations. Through initiatives like the Belt and Road Initiative (BRI) (McBride, 2023), China aims to support and invest in sustainable infrastructure projects in partner countries, fostering a global environment conducive to low-carbon development. Despite its rapid economic growth and complex challenges, China's unwavering commitment to ambitious emissions targets and sustainable development reflects a clear recognition of its global responsibilities. By forging ahead with innovative policies and investments in clean energy, China is not only contributing to the collective effort in addressing climate change but is also setting an example for other nations to follow suit. As the world's largest emitter of greenhouse gases, China's leadership in emissions reduction is crucial in shaping the trajectory of global climate action (McBride, 2023).

2.5 *Insights from South Africa*

South Africa, as a prominent member of the BRICS group, has recognized the urgency of addressing climate change vulnerabilities and has adopted a multifaceted approach that centres on renewable energy promotion and bolstering adaptation strategies. As a nation particularly susceptible to the impacts of climate change, South Africa is proactively implementing policies aimed at mitigating environmental risks and building resilience to ensure a sustainable and secure future. One of the key pillars of South Africa's climate change response is its emphasis on renewable energy adoption. The country has strategically harnessed its abundant renewable resources, such as solar, wind, and biomass, to diversify its energy mix and reduce its reliance on fossil fuels. This move not only contributes to curbing greenhouse gas emissions but also enhances energy security and reduces the nation's carbon footprint (Su, 2020).

Su et al. (2020), alluded the Renewable Energy Independent Power Producer Procurement Programme (REIPPPP) is a flagship initiative that showcases South Africa's commitment to renewable energy. Through this program, the country has successfully attracted private investment in renewable energy projects, fostering a flourishing renewable energy market. This has resulted in the addition of significant renewable energy capacity to the national grid, furthering South Africa's journey toward a low-carbon and sustainable energy future. In addition to promoting renewable energy, South Africa recognizes the need to adapt to the changing climate and its potential impacts on various sectors.

Through adaptive measures, South Africa seeks to strengthen its ability to cope with extreme weather events, rising sea levels, and changing precipitation patterns. The implementation of climate-smart agricultural practices, water conservation measures, and sustainable infrastructure development are integral components of the country's adaptation efforts. South Africa's commitment to adaptation extends to empowering vulnerable communities to become more resilient in the face of climate-related challenges. The country is actively engaged in community-based adaptation projects that focus on enhancing local capacities to cope with climate-induced impacts. This community-centred approach ensures that the most vulnerable populations are equipped with the tools and knowledge to withstand the effects of a changing climate (Mahalik, 2023).

2.6 The Reality of Climate Change—South Africa

The integration of renewable energy promotion and adaptation strategies in South Africa's climate response plan highlights the country's holistic approach to tackling climate change. By investing in clean energy solutions and enhancing adaptive capacities, South Africa is positioning itself to address both the causes and consequences of climate change. South Africa's actions in renewable energy and adaptation serve as a powerful example for other nations, particularly within the BRICS group, to follow suit (Climate Action Tracker, 2023). The country's proactive approach in embracing sustainable practices not only contributes to global climate efforts but also enhances its position as a responsible global actor committed to fostering a greener and more resilient future for its citizens and the world at large. Through these concerted efforts, South Africa continues to pave the way for sustainable development and environmental stewardship within the BRICS economies and beyond.

The policies regarding climate change and sustainable development in the BRICS countries (Brazil, Russia, India, China, and South Africa) exhibit a mix of similarities and differences due to their distinct economic, social, and environmental contexts. A comparative analysis of these policies reveals various approaches and priorities among the nations.

Table 1 provides a comprehensive comparison of climate change and sustainable development policies in the BRICS countries (Brazil, Russia, India, China, and South Africa) and includes a column highlighting events that have generated climate change in the region. The first section of the table outlines the emission reduction targets set by each country, showcasing their level of commitment and ambition in contributing to global efforts to combat climate change. The second section compares the extent to which each BRICS nation is promoting renewable energy sources, such as solar, wind, hydro, and biomass. This demonstrates their dedication to transitioning to cleaner and more sustainable energy alternatives (Climate Action Tracker, 2023).

Moving on, the table highlights the adaptation measures and strategies implemented by each country in response to climate change impacts. It showcases their efforts to protect vulnerable communities and sectors from extreme weather events and environmental changes. Another critical aspect examined in the table is the integration of climate change and sustainable development policies into the broader national development plans of each BRICS country. This indicates their commitment to harmonizing economic, social, and environmental goals for a more holistic and sustainable approach. In addition to exploring the progress and strengths of each country's policies, the table also identifies major challenges and barriers they face in effectively implementing their climate change and sustainable development initiatives. These challenges may vary depending on unique economic, political, and social conditions present

Table 1 Comparison of climate change and sustainable development policies in BRICS countries and events generating climate change in the region, Moosa (2023)

Country	*Climate change and sustainable development policies*	*Events that have generated Climate Change in the Region*
Brazil	Commendable efforts in reducing deforestation rates and promoting renewable energy sources	– Deforestation in the Amazon rainforest due to agricultural expansion and logging. – Carbon emissions from land-use change and forest degradation
Russia	– Challenges in reducing emissions due to heavy dependence on fossil fuels, especially oil, natural gas, and coal	–Significant carbon emissions from fossil fuel combustion and industrial processes. – Methane emissions from oil and gas operations and permafrost thawing
India	– Exemplary in adopting renewable energy technologies. – Global leader in solar power deployment. – Ambitious targets through initiatives like JNNSM. – Proactive in attracting investments and implementing supportive policies	–High carbon emissions from rapid industrialization and fossil fuel-based energy. – Air pollution from transportation and coal-fired power plants
China	– Commitment to ambitious emissions reduction targets. – Goal of peaking carbon dioxide emissions by 2030 and achieving carbon neutrality by 2060. – Significant progress in renewable energy deployment, particularly in solar and wind power. – Focus on energy efficiency and transitioning away from coal-fired power plants	–Extensive carbon emissions from coal-fired power plants and heavy industries. – Air pollution from rapid urbanization and transportation growth
South Africa	– Focus on renewable energy promotion and adaptation strategies. – Successful initiatives like REIPPPP in attracting private investments in renewables. – Emphasis on building resilience in sectors like agriculture, water resources, and infrastructure to cope with climate-related risks. – Acknowledging vulnerability to climate change impacts	– Greenhouse gas emissions from coal-based electricity generation and mining activities. – Water scarcity and droughts affecting agriculture and water resources

in each nation. Finally, the table includes a column listing significant events that have contributed to climate change in the region. These events may include extreme weather phenomena, deforestation, industrial activities, or other human-induced factors that have impacted the climate. By presenting this comprehensive comparison, the table offers valuable insights into the varying approaches and progress of the BRICS countries in addressing climate change and fostering sustainable development. It allows policymakers, researchers, and stakeholders to identify areas of convergence, share best practices, and collaborate on global initiatives to collectively combat climate change and create a more sustainable future for the region and the world.

2.6.1 Comparative Analysis

The climate change and sustainable development policies of the BRICS countries (Brazil, Russia, India, China, and South Africa) exhibit both similarities and differences. While all nations recognize the importance of addressing these global challenges, their approaches vary due to unique economic, social, and environmental contexts (UN-Habitat, 2023). Brazil is lauded for reducing deforestation and promoting renewable energy, while Russia struggles with emissions reduction due to its heavy reliance on fossil fuels. India excels in adopting renewable energy, and China commits to ambitious emissions targets. South Africa focuses on renewable energy and adaptation strategies to tackle climate change vulnerabilities. While all BRICS nations share common goals, their policy strengths and challenges require tailored approaches to achieve effective climate action and sustainable development.

Table 2 compares the climate change and sustainable development policies of the BRICS countries (Brazil, Russia, India, China, and South Africa) across different areas of comparison. For each country, it highlights their approach in promoting renewable energy, aligning with sustainable development goals, mitigating climate change, implementing climate adaptation strategies, and engaging in global cooperation. The '✔' indicates that a country is actively involved in that area, while the '✘' indicates challenges or limitations in their approach. This table provides a concise and comprehensive overview of the similarities and differences among the BRICS countries in their efforts to address climate change and promote sustainable development.

Table 2 Comparison of climate change and sustainable development policies among BRICS countries, Moosa (2023)

Area of Comparison	*Brazil*	*Russia*	*India*	*China*	*South Africa*
Renewable Energy Promotion	✔ Promotes renewable energy sources such as solar, wind, and bioenergy	✘ Lags in adopting renewable energy due to heavy reliance on fossil fuels	✔ Actively embraces solar and wind power, aiming for 175 GW of renewable capacity by 2022	✔ Leads in global solar and wind power installations and invests significantly in renewables	✔ Implements the Renewable Energy Independent Power Producer Procurement Programme (REIPPPP) to attract private investments in renewables
Sustainable Development Goals	✔ Aligned with UN SDGs, emphasizing inclusive growth, poverty eradication, and environmental protection	✔ Committed to SDGs, with emphasis on economic development, but faces challenges in social equity	✔ Strong commitment to achieving SDGs, focusing on social inclusion, education, and clean water	✔ Prioritizes SDGs, particularly on poverty alleviation, clean energy, and sustainable cities	✔ Integrates SDGs into the National Development Plan (NDP), emphasizing economic, social, and environmental dimensions
Climate Change Mitigation	✔ Successfully reduced deforestation rates and supports renewable energy adoption	✘ Struggles to reduce emissions due to heavy reliance on fossil fuels	✔ Making progress in emissions reduction, transitioning to cleaner energy sources	✔ Commits to ambitious emissions reduction targets and actively adopts renewable energy	✔ Implements policies to reduce emissions and transition to a low-carbon economy

(continued)

Table 2 (continued)

Area of Comparison	*Brazil*	*Russia*	*India*	*China*	*South Africa*
Climate Adaptation	✔ Implements adaptation strategies to tackle climate impacts, especially in vulnerable regions	✘ Faces challenges in climate adaptation due to vast landmass and varying ecosystems	✘ Focuses less on adaptation due to economic priorities, but addresses regional vulnerabilities	✘ Prioritizes mitigation, but takes measures to build resilience against climate impacts	✔ Implements adaptation strategies, particularly in sectors like agriculture and water management
Global Cooperation	✔ Actively engages in international climate agreements and shares best practices	✔ Collaborates with other nations on climate initiatives, but with varying levels of ambition	✔ Engages in global partnerships and shares expertize, particularly in renewable energy	✔ Takes a leading role in international climate negotiations and cooperation	✔ Engages in international collaborations and community-based adaptation projects

2.6.2 *Swot Analysis: South Africa's Climate Change and Sustainability Efforts*

The hypothesis for the SWOT analysis is that South Africa possesses significant strengths in its renewable energy potential, renewable energy progress, and adaptation strategies. However, the country faces challenges due to limited financial resources and technological capacity, which could hinder the rapid scaling up of sustainable initiatives (Table 3).

2.6.3 *Challenges and Barriers in South Africa's Climate Change Mitigation and Sustainable Development Within the BRICS Context*

Challenges and barriers that South Africa encounters in its pursuit of climate change mitigation and sustainable development within the BRICS context reveals several key findings. South Africa faces economic challenges that impact its ability to invest in renewable energy infrastructure and technology. The country's energy sector is heavily reliant

Table 3 SWOT analysis: South Africa's climate change and sustainability efforts, Moosa (2023).

Strengths	*Weaknesses*	*Opportunities*	*Threats*
1. Abundant Renewable Resources: Solar, wind, and biomass resources offer clean energy potential	1. Limited Financial Resources: Financial constraints hinder rapid implementation	1. International Cooperation: Collaborating with BRICS nations offers shared learning and support	1. Climate Vulnerabilities: Droughts, heatwaves, and extreme events pose risks to sectors and well-being
2. Renewable Energy Progress: Successful initiatives like REIPPPP attract private investments	2. Technology Capacity: R&D in renewable energy may require further strengthening	2. Global Partnerships: Engaging in initiatives attracts investment and expertize	2. Global Policy Shifts: Changes in international policies may impact finance and technology support
3. Adaptation Strategies: Comprehensive plans to tackle climate impacts in vulnerable sectors		3. Community-Based Adaptation: Empowering vulnerable communities enhances resilience	3. Geopolitical Challenges: Balancing climate commitments with economic interests presents challenges
4. National Climate Policies: NCCRP sets a strong foundation for climate action			

on coal, which poses a significant barrier to achieving emissions reduction targets. Additionally, economic constraints limit the government's capacity to implement large-scale renewable energy projects and incentivize private sector investment. South Africa encounters social challenges related to energy access and affordability. While transitioning to renewable energy sources is essential for sustainable development, it must be done inclusively to ensure that vulnerable communities are not left behind (Mechanic & Tanner, 2007). Access to affordable clean energy remains a concern, particularly in rural areas, where grid connectivity is limited.

Policy and regulatory challenges hinder the progress of climate change mitigation and sustainable development in South Africa. Inconsistent policy frameworks and bureaucratic barriers can create uncertainty for investors and hinder the effective implementation of sustainable initiatives. Harmonizing policies within the BRICS group is essential for promoting collaborative efforts and avoiding policy conflicts.

South Africa faces technological challenges in adopting and integrating renewable energy technologies into its energy mix. The country must enhance its technological capabilities to effectively harness the potential of solar, wind, and other renewable sources. Geopolitical factors also present challenges within the BRICS context. Each member country has unique national interests and priorities, making it challenging to align efforts and coordinate climate change mitigation strategies. Geopolitical tensions and trade relations can impact the willingness of BRICS nations to collaborate effectively (Liu, 2016).

South Africa encounters environmental challenges, such as water scarcity and climate-induced impacts like extreme weather events, these challenges amplify the urgency for sustainable development and necessitate adaptive strategies to build resilience to climate change (Development Bank of South Africa, 2023). Each member country of the BRICS group has unique national interests and priorities that can influence their approach to climate change and sustainable development (Table 4).

The findings presented herewith allude to the need for policy synthesis and actual joining of hands in the BRICS bloc to motivate excellent partnerships which garner continued development directed towards a sustainable future for all member countries. This chapter makes recommendation for collaborative practices to become defined and structured as follows in Table 5.

3 Recommendations and Conclusion

The research conducted on the climate change and sustainable development policies of the BRICS countries, with an emphasis on South Africa's perspective, provides valuable insights. However, there are several areas that warrant further investigation and research to deepen our understanding and inform future policy decisions. Future research on climate change and sustainable development within the BRICS countries should consider several key areas. First, longitudinal studies are needed to assess

Table 4 Key areas of national interest and sustainable development challenges in BRICS countries, Moosa (2023)

Country	*Key areas of national interests*	*Priorities for sustainable development*	*Specific challenges*
Brazil	Renewable energy promotion, sustainable development goals, climate change mitigation, climate adaptation	Promoting renewable energy sources, reducing poverty and inequality, protecting the Amazon rainforest	Deforestation, air pollution, water scarcity
Russia	Sustainable development goals, climate change mitigation, global cooperation	Reducing emissions, transitioning to a low-carbon economy, cooperating with other countries on climate change	Heavy reliance on fossil fuels, permafrost thawing, air pollution
India	Renewable energy promotion, sustainable development goals, climate change mitigation	Increasing access to clean energy, reducing poverty and inequality, improving water quality	High carbon emissions, air pollution, water scarcity
China	Sustainable development goals, climate change mitigation, global cooperation	Peaking carbon dioxide emissions by 2030 and achieving carbon neutrality by 2060, transitioning to a low-carbon economy, cooperating with other countries on climate change	Extensive carbon emissions, air pollution, water scarcity
South Africa	Renewable energy promotion, sustainable development goals, climate change adaptation	Increasing access to clean energy, reducing poverty and inequality, building resilience to climate change	Greenhouse gas emissions from coal-based electricity generation and mining activities, water scarcity and droughts

the long-term impact of policies in each country, evaluating their effectiveness in reducing emissions, preserving the environment, and fostering socio-economic development. Second, research should explore the social equity implications of climate change policies, focusing on how vulnerable and marginalized communities are affected and identifying ways to promote climate justice. Third, the role of technological innovation and potential for technology transfer and collaboration among BRICS nations should be examined, particularly in the areas of renewable energy, energy

Table 5 Strategies and policy recommendations for South Africa's harmonization with BRICS countries in addressing climate change and advancing sustainable development, Moosa (2023)

Joint policy recommendation	*Illustrative action*
Increased Collaboration on Research and Development	Establish a BRICS Research and Development (R&D) consortium focused on climate change and sustainable development. Pool financial and intellectual resources to conduct joint research projects, share data, and develop innovative technologies, such as advanced solar panels, energy storage systems, and climate-resilient agricultural practices. Facilitate knowledge exchange through regular workshops, conferences, and technology transfer programs. Foster collaboration between research institutions, universities, and private enterprises to accelerate the development and deployment of cutting-edge solutions
Joint Investments in Sustainable Infrastructure	Formulate a BRICS Sustainable Infrastructure Investment Fund to finance joint projects aimed at reducing carbon emissions and promoting sustainable urbanization. Prioritize investments in renewable energy infrastructure, such as large-scale solar and wind farms, as well as modernizing transportation systems with electric vehicles, high-speed rail networks, and smart urban planning. Develop eco-friendly water management solutions, including wastewater treatment facilities and desalination plants. Implement green building standards and retrofit existing structures for energy efficiency. Enhance connectivity between BRICS countries through cross-border infrastructure projects to facilitate sustainable trade and economic growth

(continued)

Table 5 (continued)

Joint policy recommendation	*Illustrative action*
Knowledge Sharing and Best Practices Exchange	Create a BRICS Climate and Sustainable Development Knowledge Platform, accessible to policymakers, researchers, and civil society. Share best practices and success stories on climate change mitigation and adaptation efforts. Collaborate on capacity-building programs to train experts in sustainable development planning and implementation. Organize joint workshops and chapter tours to promote peer learning and experience sharing. Establish a digital repository of policies and strategies to support evidence-based decision-making and policy formulation. Encourage regular policy dialogues and expert consultations to strengthen cooperation on climate and sustainability challenges
Coordinated Advocacy at International Forums	Establish a BRICS Climate Diplomacy Task Force to coordinate positions and messages at international climate conferences, such as COP negotiations. Develop a shared agenda for advancing the Paris Agreement's goals and advocate for increased climate finance and technology transfer to support developing nations. Promote a unified voice on climate justice and equitable burden-sharing among developed and developing economies. Collaborate on joint initiatives to address global challenges, such as deforestation, biodiversity loss, and ocean conservation. Engage with other major economies and international organizations to strengthen multilateral climate action and promote inclusive, sustainable development

(continued)

Table 5 (continued)

Joint policy recommendation	*Illustrative action*
Develop a National Climate Change Adaptation Plan	Conduct a comprehensive vulnerability assessment to identify climate risks and impacts across different regions and sectors in South Africa. Formulate a National Adaptation Plan (NAP) that prioritizes climate resilience in critical areas, including water resources, agriculture, coastal zones, and human settlements. Engage stakeholders from government, civil society, and private sectors to ensure inclusivity and ownership of the adaptation plan. Implement nature-based solutions, such as reforestation and ecosystem restoration, to enhance resilience and ecosystem services. Establish early warning systems and disaster preparedness measures to respond to extreme weather events and climate-related emergencies
Set Ambitious Emissions Reduction Targets	Develop a long-term Low-Emission Development Strategy (LEDS) that outlines a clear pathway to achieving net-zero emissions by mid-century. Set intermediate emissions reduction targets in alignment with global temperature goals to ensure South Africa's fair contribution to the global effort. Implement sector-specific policies and regulations to decarbonize energy, industry, transportation, and agriculture. Phase out coal-fired power plants and transition to cleaner energy sources while ensuring a just transition for affected communities and workers. Encourage private sector involvement through carbon pricing mechanisms and emissions trading systems. Monitor and report progress regularly to enhance transparency and accountability in emissions reduction efforts

(continued)

Table 5 (continued)

Joint policy recommendation	*Illustrative action*
Invest in Renewable Energy	Introduce favourable policies and incentives to attract domestic and foreign investments in renewable energy projects. Establish feed-in tariffs, tax incentives, and green bonds to finance the development of solar, wind, and hydropower installations. Promote public–private partnerships to accelerate the deployment of renewable energy technologies. Invest in research and innovation to enhance the efficiency and cost-effectiveness of renewable energy systems. Build a skilled workforce capable of supporting the renewable energy sector's growth through training and education programs. Ensure the integration of renewable energy into the national grid for reliable and resilient power supply
Promote Energy Efficiency	Enforce energy efficiency standards for buildings, appliances, and industrial processes. Provide financial incentives, such as energy efficiency subsidies and tax breaks, to encourage businesses and households to adopt energy-saving technologies. Launch public awareness campaigns to promote energy conservation and behavioural change. Strengthen building codes and standards to ensure new constructions meet high energy performance criteria. Develop smart cities and sustainable urban planning to optimize energy use and reduce emissions. Support research and development of innovative energy-efficient technologies and practices. Establish partnerships with industries to implement energy management systems and reduce their carbon footprint

efficiency, and green technologies. Fourth, a detailed analysis of financing mechanisms is essential, including the effectiveness of public and private sector funding, international assistance, and innovative instruments in mobilizing climate-related resources. Fifth, the impact of climate-induced migration and the adaptation strategies required to address its consequences warrant investigation. Sixth, studies should examine public perception, awareness, and attitudes towards climate change and sustainable development, identifying communication strategies that can influence behavioural change. Seventh, the contributions of the private sector and civil society to climate action and sustainability efforts should be analyzed, highlighting successful partnerships and initiatives. Lastly, research should evaluate the BRICS countries' climate diplomacy efforts and explore opportunities for stronger international cooperation and a unified BRICS stance in global climate negotiations. By conducting research in these areas, policymakers, academics, and stakeholders can gain deeper insights into the complexities of climate change and sustainable development within the BRICS context. This knowledge can inform evidence-based policy decisions, foster international collaboration, and accelerate progress towards a more sustainable and climate-resilient future for the BRICS countries and the world.

3.1 Conclusion

This chapter explored the climate change and sustainable development policies of the BRICS countries, with a focus on South Africa's perspective within the group. Through comparative and SWOT analyses, common goals, strengths, and challenges were assessed, paving the way for policy recommendations to align South Africa's approach with other BRICS nations. The analysis revealed that while each country has unique national interests and priorities, they also share common objectives in combating climate change and promoting sustainable development. By fostering greater collaboration on research, investments in sustainable infrastructure, and sharing best practices, the BRICS nations can collectively enhance their efforts and influence global environmental action positively. The policy recommendations put forward aim to support South Africa in harmonizing its approach with other BRICS countries, fostering cooperation, and achieving ambitious climate and sustainability goals. They stress the importance of developing a national climate adaptation plan, setting aggressive emissions reduction targets, and increasing investments in renewable energy and energy efficiency. Promoting social equity and climate justice and exploring technological innovation and transfer will be essential for inclusive and sustainable development.

The findings from this chapter contribute to the broader discussion on climate change and sustainable development, emphasizing the significance of international cooperation and collaboration in addressing global environmental challenges. The recommendations offer valuable guidance to policymakers and stakeholders in the BRICS nations, enabling them to enhance their collective efforts towards a more sustainable and resilient future (Santana et al., 2014). By acting upon these suggestions and continuing to explore avenues for innovation and knowledge exchange, the BRICS countries can collectively contribute significantly to global climate change mitigation and sustainable development.

References

Climate Action Tracker. (2023). *Countries climate action rating*. CAT.

Crompton, P., & Wu, Y. (2005). Energy consumption in China: Past trends and future directions. *Energy Economics, 27*(1), 195–208. Retrieved July 31, 2023, from http://science.uwa.edu.au/__data/assets/pdf_file/0003/102567/04_22_crompton_wu.pdf

Development Bank of South Africa. (2023). *The effects of poor infrastructure in education, transport and communities*. DBSA.

Development, W. C. (n.d.). *Our common future, Chapter 2: Towards sustainable development—A/42/427 Annex, Chapter 2—UN documents: Gathering a body of global agreements*. Retrieved July 31, 2023, from http://www.un-documents.net/ocf-02.htm

Global Ocean Forum. (2010). *Oceans day at Durban*. Global Ocean Forum.

Grantham Research Institute. (2011). *National Climate Change Response Policy White Paper (NCCRP)*. Climate Change Laws of the World.

Grantham Research Institute on Climate Change and the Environment. (2022). *Brazil First NDC—Second update*. Brazil's Nationally Determined Contributions.

Hau, M. v., Scott, J., & Hulme, D. (2012). Beyond the BRICs: Alternative strategies of influence in the global politics of development. *The European Journal of Development Research, 24*(2), 187–204. Retrieved July 30, 2023, from https://link.springer.com/article/10.1057/ejdr.2012.6

Haywood, J. (2016). *Climate change: Observed impacts on planet earth* (T. M. Letcher, Ed.) Elsevier. Retrieved July 30, 2023.

IEA. (2012). *Key world energy statistics*. Retrieved 7 31, 2023, from http://www.iea.org/publications/freepublications/publication/kwes.pdf

India Meteorological Department. (2023). Retrieved July 31, 2023, from http://www.imd.gov.in

International Energy Agency. (2021). *Jawaharlal Nehru National Solar Mission (Phase I, II and III)*. IEA.

International Monetary Fund. (2023). *GDP of BRICS countries 2000 to 2028*. Statista.

Kıprızlı, G. a. (2023). The onset of BRICS cooperation on climate change: material change, ideational convergence and the road to Copenhagen 2009. *Third World Quarterly*, 1–19.

Lebedev, F. (2022). *Russia's first climate lawsuit filed over greenhouse emissions*. Reuters.

Liu, D. (2016). China's sponge cities to soak up rainwater. *Nature, 537*(7620), 307–307. Retrieved April 1, 2023, from https://nature.com/articles/537307c

Mahalik, M. M. (2021). Do educational levels influence the environmental quality? The role of renewable and non-renewable energy demand in selected BRICS countries with a new policy perspective. *Renewable Energy*, 419–432.

Mahalik, M. M. (2023). Do educational levels influence the environmental quality? The role of renewable and non-renewable energy demand in selected BRICS countries with a new policy perspective. *Renewable Energy*, 419–432.

McBride, J. (2023). *China's massive belt and road initiative*. Council on Foreign Relations.

Mechanic, D., & Tanner, J. K. (2007). Vulnerable people, groups, and populations: Societal view. *Health Affairs, 26*(5), 1220–1230. Retrieved July 31, 2023, from https://healthaffairs.org/doi/full/10.1377/hlthaff.26.5.1220

Meena, K. (2013). BRICS: an explanation in critical geography. *Contexto Internacional, 35*(2), 565–593. Retrieved July 30, 2023, from http://scielo.br/scielo.php?script=sci_arttext&pid=s0102-85292013000200009

Myhre, G. D.-M.-F. (2013). *Climate Change 2013: The physical science basis. contribution of working group I to the Fifth Assessment Report of the Intergovernmental Panel on Climate Change*. (T. F. Stocker, D. Qin, G.-K. Plattner, M. Tignor, S. K. Allen, J. Boschung,... P. M. Midgley, Eds.). Cambridge University Press. Retrieved July 30, 2023, from http://www.ipcc.ch/report/ar5/wg1

O'Neill, J. (2001). *Building Better Global Economic BRICs, Global Economics Paper No: 66. (66)*. Retrieved July 30, 2023, from http://www.goldmansachs.com/our-thinking/archive/archive-pdfs/build-better-brics.pdf

Santana, N. B., Rebelatto, D. A., Périco, A. E., & Mariano, E. B. (2014). Sustainable development in the BRICS countries: An efficiency analysis by data envelopment. *International Journal of Sustainable Development and World Ecology, 21*(3), 259–272. Retrieved July 30, 2023, from https://tandfonline.com/doi/full/10.1080/13504509.2014.900831

Su, W. W. (2020). arbon dioxide emission decomposition along the gradient of economic development: The case of energy sustainability in the G7 and Brazil, Russia, India, China and South Africa. *Sustainable Development*, 657–669.

Transforming our world: the 2030 Agenda for Sustainable Development. (2023). Retrieved July 31, 2023, from https://sustainabledevelopment.un.org/post2015/transformingourworld

UN Climate Summit. (2014). *UN Climate Summit*. Retrieved July 13, 2023, from https://www.un.org/climatechange/summit/

UN-Habitat. (2023). Retrieved June 13, 2023, from https://unhabitat.org/urban-knowledge/awards/

Ventura, R. F. (2023). *Brics inbound and outbound tourism versus socio-economic indicators*. GeoJournal of Tourism and Geosites.

Zhou, N. (2011). China's Energy and Carbon Emissions Outlook to 2050. *Lawrence Berkeley National Laboratory*. Retrieved July 31, 2023, from https://china.lbl.gov/sites/all/files/lbl-4472e-energy-2050april-2011.pdf

CHAPTER 6

Socioeconomic Impacts of Climate-Driven Disasters: The South African Perspective

Simphiwe Emmanuel Dube and Zikho Qwatekana

1 Introduction

South Africa is increasingly exposed to disaster risks that impact both human and natural systems. These include extreme weather events, rising urbanization, fragile infrastructure, land degradation, pandemics, and socioeconomic vulnerabilities. The April 2023 floods in KwaZulu-Natal exemplify how disasters damage critical infrastructure such as schools, hospitals, roads, and fire stations while also disrupting livelihoods and deepening inequality. Research shows that communities with lower socioeconomic status (SES) suffer disproportionate impacts from

S. E. Dube · Z. Qwatekana (✉)
Department of Tourism, Nelson Mandela University, Port Elizabeth, South Africa
e-mail: zikhoq@dut.ac.za

Department of Ecotourism, Durban University of Technology, Durban, South Africa

S. E. Dube
e-mail: simphiwe.dube@mandela.ac.za

N. Tshikovhi et al. (eds.), *The BRICS Imperative*,
https://doi.org/10.1007/978-3-032-05932-1_6

disasters, including greater losses of property, income, and physical well-being (SAMHSA, 2017a). According to Hallegatte et al. (2017), disasters reinforce poverty cycles by stripping vulnerable groups of assets and preventing recovery. This raises concern about long-term resilience in already fragile communities. Disaster risk is no longer just a humanitarian issue; it is now a development crisis affecting human capital, economic productivity, and social stability (IFRC, 2022; UNDRR, 2019). Despite global awareness of these challenges, a key knowledge gap persists in understanding how physical infrastructure damage from disasters is causally and systematically linked to social and economic consequences. Existing literature tends to separate physical assessments such as, damage to buildings and roads from socioeconomic indicators like income loss or migration, making it difficult for policymakers to plan holistic recovery strategies. The need to bridge this gap is urgent, particularly in the Global South, where disasters are increasing in frequency and intensity, yet recovery planning is often fragmented or underfunded (Lavell et al., 2012; Pelling & Wisner, 2009). This chapter draws from disaster risk reduction (DRR) and resilience theory to propose a framework that connects physical disaster damage to social and economic outcomes. The conceptual basis includes the pressure-and-release model (Wisner et al., 2004), which highlights vulnerability and exposure as key mediating factors, and systems theory, which views disasters as multidimensional shocks disrupting complex social-ecological systems.

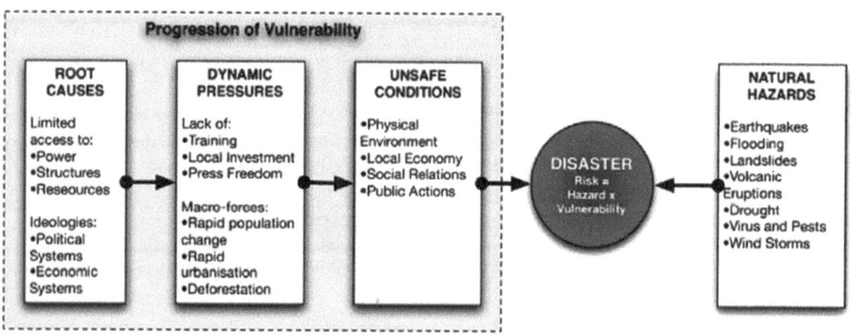

Pressure and Release model (Wisner et al., 2004: 51)

The model suggests that disasters are not purely natural phenomena, they result from the intersection of environmental hazards, some which are climate driven, and structurally embedded vulnerabilities. The Pressure and Release (PAR) model reframes disasters as the product of Hazard plus Vulnerability, emphasizing that their impacts are profoundly shaped by social, economic, and political factors.

1.1 Aim and Objectives

This chapter aims to investigate the link between disaster-induced physical infrastructure damage and its broader social and economic impacts in South Africa.

The specific objectives are to:

1. Explore existing literature on the socioeconomic impacts of disaster-related infrastructure damage.
2. Identify the key factors that mediate the relationship between physical damage and social vulnerability.
3. Recommend evidence-informed strategies for post-disaster economic recovery and resilience-building in affected communities.

1.2 The Concepts of Disaster and Socioeconomic Disaster

Disasters are commonly defined as serious disruptions to the functioning of a community or society that exceed its capacity to cope using its own resources (IFRC, 2022). These events result in significant human, material, economic, and environmental losses. Disasters can stem from natural hazards such as floods, earthquakes, droughts, technological accidents including industrial explosions, cyberattacks, or human-induced crises. What transforms a hazard into a disaster is the degree of exposure and vulnerability of the affected population factors often linked to poverty, inequality, weak infrastructure, and institutional fragility (Cutter et al., 2003; UNDRR, 2019).

The concept of a socioeconomic disaster refers to the cascading effects of such disruptions on the economic and social fabric of communities. These impacts go beyond immediate physical destruction, often triggering long-term consequences such as unemployment, homelessness, reduced access to education and healthcare, food insecurity, and a breakdown in social cohesion (Lavell et al., 2012; Wisner et al., 2004). Socioeconomic disasters often deepen pre-existing inequalities and disproportionately affect marginalized groups, especially those with limited coping mechanisms and access to recovery resources (Hallegatte et al., 2017). One of the most profound and far-reaching disasters in living memory is the COVID-19 pandemic, which exemplifies both physical and socioeconomic disruption on a global scale. According to the World Health Organization (WHO), as of October 2022, COVID-19 had claimed over 6.5 million lives globally and infected more than 630 million people (WHO, 2022). Its effects were not limited to health; the pandemic caused widespread economic instability, education disruption, and mental health deterioration. In South Africa, the socioeconomic effects of the pandemic were particularly pronounced. As of April 2023, the country recorded over 4,076,463 confirmed cases and 102,595 deaths due to COVID-19, with 3,912,506 recoveries (NICD, 2023). Beyond these statistics, the pandemic exacerbated unemployment, strained public health services, and disrupted schooling for millions of learners. A report by Stats SA (2021) noted that more than 2.2 million jobs were lost during the first year of the pandemic, pushing the unemployment rate above 34%. Vulnerable populations, particularly in informal settlements and rural areas, experienced intensified food insecurity, loss of income, and inadequate access to healthcare and social support. Disasters like COVID-19 highlight the interconnectedness of health, governance, infrastructure, and economic resilience. They underscore the importance of integrating socioeconomic dimensions into disaster preparedness and response strategies particularly in high-risk countries like South Africa where systemic vulnerabilities amplify disaster impacts. The economic impacts of natural disasters are multifaceted, encompassing both direct and indirect (macro-level) consequences. Direct economic impacts arise when disasters damage tangible assets such as buildings, roads, power infrastructure, and productive capital. For instance, the 2010 earthquake in Haiti destroyed over 105,000 homes and damaged another 188,000 (PDNA, 2010), while the 2019 Cyclone Idai in Mozambique led to the destruction of health facilities, schools, and transport infrastructure, causing estimated losses exceeding USD 1 billion (UNDRR, 2019).

These direct losses typically occur at or shortly after the disaster site and may include the destruction of homes, businesses, crops, livestock, and community infrastructure—as well as the deterioration of physical and emotional wellbeing. Catastrophe modelling and empirical damage assessments are commonly used to quantify these losses (Koks et al., 2019; Sapountzaki, 2019). Indirect impacts, on the other hand, refer to economic disruptions that occur after the initial event as a consequence of the direct effects. These include interruptions to economic activity, shifts in employment patterns, and strains on public finances. In some cases, positive secondary effects may arise, such as temporary economic boosts from reconstruction efforts. For example, post-disaster recovery in New Orleans after Hurricane Katrina showed mixed trends in regional economic activity (Cavallo & Noy, 2011). These indirect effects often manifest through declines in household income, trade, investment, and productivity. They are typically modelled using macroeconomic frameworks and are referred to as higher-order effects. Indicators such as GDP growth, employment, consumption, and inflation are commonly used to assess these impacts (Botzen et al., 2019; Tasri et al., 2022).There is a significant correlation between pre-existing economic vulnerability and disaster outcomes. Poorer households and low-income regions suffer disproportionately, not only due to weaker infrastructure but also due to limited access to insurance, social safety nets, and emergency services (Lovell & Le Masson, 2014). These socioeconomic disasters reinforce cycles of poverty and inequality, often trapping communities in long-term disadvantage. Understanding both the direct and indirect economic dimensions of disasters is crucial for building resilience and planning recovery in an inclusive and equitable manner.

2 Infrastructure Damage

Infrastructure damage is one of the most immediate and visible consequences of natural and technological disasters. Infrastructure in this context refers to essential physical systems such as energy grids, water supply, sanitation, transport, housing, communication networks, and healthcare facilities. Disasters impact these systems in ways that not only disrupt public services but also compromise economic productivity, human health, and social cohesion. Natural hazards such as floods, earthquakes, and hurricanes can damage or destroy infrastructure directly through physical force or indirectly by creating conditions that degrade these systems over time. For example, during the 2011 Great East Japan Earthquake, widespread destruction of transport systems, energy infrastructure, and industrial facilities resulted in economic losses estimated at \$360 billion, making it one of the most expensive natural disasters in history (OECD, 2017). Similarly, the 1994 Northridge Earthquake in California revealed the vulnerability of urban infrastructure, where utility service interruptions contributed to the closure of many businesses. Among the top five reasons for business closures in the aftermath, power outages and phone service failures ranked second and fourth respectively (Tierney, 1997). Disasters often highlight the fragility of infrastructure that is already under strain due to poor maintenance, ageing, or inadequate planning. The European Union (2023) emphasizes that disruptions in energy, water, health, transport, and financial services can lead to cascading failures with long-term economic and social consequences. These impacts are magnified in low-income communities where adaptive capacity is limited.

3 Health Infrastructure

Health infrastructure is frequently strained or rendered non-functional during disasters. Damage to hospitals, clinics, and medical supply chains reduces access to critical care at the exact moment it is most needed. For example, Cyclone Idai in 2019 damaged 66 health facilities in Mozambique, leaving thousands without medical care during a concurrent cholera outbreak (UNDRR, 2019). In such cases, the direct destruction of infrastructure is compounded by the disruption of services, including immunization, maternal care, and emergency response. Additionally, poor sanitation systems and lack of clean water a result of damaged water infrastructure are known contributors to morbidity and mortality following floods and cyclones (Scapini, 2020). Rather than speculate, these findings are based on field research and disaster post-impact assessments.

4 Housing and Built Environment

Housing infrastructure is especially vulnerable in informal settlements and low-income urban areas. People with lower socioeconomic status often live in high-risk locations such as, floodplains, steep slopes, and in poorly constructed buildings. During disasters, these homes are easily destroyed, leading to displacement, asset loss, and social instability. Research by SAMHSA (2017b) and Mwape (2009) confirms that poorer populations are more likely to lose their homes and belongings and are less equipped to recover. A practical South African example is the repeated flooding in Cape Town's informal settlements, where between 32,000 and 34,000 people are displaced annually during the winter rainfall season in 2007, 2008, and 2009 (Ziervogel & Smit, 2009). These events exacerbate housing insecurity and deepen poverty. Displacement also disrupts community cohesion and access to essential services, reinforcing the socioeconomic vulnerability of affected populations (French, 2018). The Asian Disaster Preparedness Center (ADPC, 2011) identifies several housing-related risk factors, including poor land-use planning, lack of disaster-resilient construction, weak building regulations, and inadequate governance. These vulnerabilities, if unaddressed, significantly increase housing losses in disaster scenarios.

5 Transportation System Disruption

Transport infrastructure is critical for economic activity, especially during and after disasters. Roads, railways, bridges, and ports enable the movement of goods, services, and emergency supplies. When disasters strike, damage to these systems can delay response times, isolate communities, and stall economic activity. Rodrigue (2020) states that transport systems are considered vital infrastructure because any disruption has ripple effects on national and regional economies. During Hurricane Katrina, severe flooding damaged roads and highways, obstructing evacuation and relief efforts, while also increasing the cost and time of goods delivery (Pregnolato et al., 2017). Ports are similarly vulnerable. Achurra-Gonzalez et al. (2019) found that even temporary disruption of port operations leads to increased transport costs, commodity spoilage, and lost revenue. Verschuur et al. (2020) further note that disruptions in maritime logistics can have long-lasting effects on supply chains and economic systems.

The economic effects from disaster disruptions to the transportation system can be substantial to communities. Moreover, to the high repair costs associated with direct damage, damage caused by disasters to transportation infrastructure can also hinder the movements of people and products. Rodrigue (2020) posits that a disruption in any transportation system components can have a substantial impact on the social and economic well-being of an entire country or region, and a result transportation is regarded as an essential infrastructure. Transportation infrastructure may suffer direct or indirect damage during extreme weather events, endangering public safety and leading to serious disruptions with serious impact for the economy and society (Pregnolato et al., 2017).

Natural disasters can have a variety of impacts on transportation networks and ports operations. A port interruption can initially limit the volume of cargo that a port can handle for a set period of time, leading to delays, depreciation of commodities, and increased transportation expenses if cargo needs to be rerouted (Achurra-Gonzalez et al., 2019). If ports sustain physical damage, their ability to function may be seriously compromised. Rebuilding port assets will be expensive, and the logistics network may be permanently disrupted. Supply chains that rely on essential transportation systems may also have significant repercussions from disruptions to these systems (Verschuur et al., 2020).

6 Financial Impact

Disasters affect the world's poorest population disproportionately more from an economic perspective: Natural disasters are estimated to have caused $92 billion in economic damages in 2015; the average yearly cost is projected to be more than $300 billion (The World Bank, 2016). These figures are helpful in that they show patterns and expenses associated with catastrophes, but they fail to indicate how disasters impact people's quality of life. All social classes of people suffer equal levels of disaster damage, the rich will suffer higher financial losses since they have more assets than the poor (Okuda & Kawasaki, 2022: 5). Nonetheless, several studies have demonstrated that the impoverished with lesser earnings are more negatively affected than the wealthy with greater damage when comparing the ratio of flood damage to income. Consequently, the extent of the financial losses may be exaggerated in comparison to the true effect. Many people lose their employments and businesses because of disasters across the globe.

South Africa has been faced with compounding disaster shocks, such as COVID-19, droughts, Floods, and social upheaval, in the past three years (The World Bank, 2022). During the course of eleven days in March 2020, South Africa announced two separate states of disaster one for the epidemic and the other for the drought. In July 2021, a nationwide state of calamity was declared due to drought, coinciding with the third wave of the COVID-19 pandemic in as many years (2018, 2020, 2021). The country's macroeconomic and budgetary situation is significantly strained as a result of the compounding shocks. President Cyril Ramaphosa proposed a rescue package worth R500 billion or around 10% of GDP, when the pandemic first broke out. In 2020, the economy shrank by 6.4%. By the end of 2021, there had been massive wildfires in the Western Cape and floods in all nine provinces. The national government calculated that the floods in January 2021 alone would cost more than R50 million. Violent social disturbance occurred in KwaZulu-Natal in July 2021; damages were estimated to be R20 billion, with an estimated R50 billion being the overall economic impact. The South African government's (GoSA) responses to these shocks contributed to the debt-to-GDP ratio rising from 57% in 2018 to 80% in 2021 (The World Bank, 2022).

7 Planning for Disasters to be Economically Prepared

The purpose of the pre-disaster economic preparation plan is to get a community ready for emergencies while keeping the business community and local economy in mind. Should a calamity strike, the plan outlines the duties and procedures for those involved in economic recovery. A community can act more swiftly and effectively after a tragedy and accelerate the recovery process if they have a plan in place beforehand. Restore Your Economy ORG (2023) suggests the following steps for the pre-disaster economic preparation plan. Any pre-existing disaster plans should be supplemented and informed by the pre-disaster economic preparedness planning steps that are outlined and discussed below.

Step 1: Identify a Lead Economic Development Organization
The creation of a pre-disaster economic preparedness strategy necessitates a thorough and meticulous procedure overseen by a specialized organization capable of uniting all relevant parties. In order to launch the project, this group will need to organize the resources required and mobilize broad support for it. A more extensive, all-encompassing pre-disaster preparedness plan should eventually incorporate and integrate with the economic readiness plan.

Step 2: Hold a Kick-Off Meeting with all Economic Recovery Stakeholders
Step 2a: Create Stakeholder Groups

This step involves a wide range of active and diverse stakeholders, such as elected officials, business and public figures, and other significant community players. It will be crucial to have representation from the sectors of the community's economy. This stakeholder group may also function as the economic recovery team following the tragedy. Assemble the stakeholders into smaller groups to concentrate on particular issues that the plan will address. Making plans for what to do after a calamity can be difficult, contentious, and complex. At the kick-off meeting, all parties involved should be informed of this and agree to proceed with the decision-making process.

Step 2b: Establish Clear Goals for the Planning Process
The responsible disaster team needs to ensure that the planning process receives enough time. They need to establish a regular meeting schedule for the group at large and for its subgroups. Give each stage a defined timeframe. This will guarantee that the process proceeds, keeping the interested parties focused and informing the general public that, despite the process's duration, it is ultimately coming to an end.

Step 3: Think Strategically About How a Disaster Could Affect Economic Development
Step 3a: Identify the Community's Economic Assets
The community will be able to identify its economic vulnerabilities and the resources to be used in the case of a disaster with the aid of an inventory of present assets and economic drivers. Economic analyses and strategy plans that are currently in place may contain up-to-date information about this endeavour. A solid grasp of the following should be included in the assessment:

- Demographics;
- Important sectors and/or clusters;
- Features of labour and the workforce;
- SWOT Analysis: (Strengths, Weaknesses, Opportunities, Threats); and

- Local Policy Framework (such as construction codes, state, and local laws and incentives, policies for environmental mitigation, etc.)

Step 3b: Conduct an Economic Vulnerability Study
An Economic Vulnerability Analysis Study needs to be conducted for the specific type of disaster. The preceding step's economic analysis ought to be synchronized with the findings of hazard mitigation and land-use planners in order to determine which businesses, sectors, or industry clusters are more susceptible to disasters.

Step 3c: Conduct Scenario Planning for Community Redevelopment
Analyze the economic resources and cluster industries of the town in relation to potential disaster scenarios.
How to Plan a Scenario

- Describe the disaster's backdrop, including the type of disaster the town is experiencing;
- Determine potential community impact variables by guiding important stakeholders through role-playing activities;
- Find out from the stakeholders which economic and communal assets in the area are most at risk from these variables;
- Combine all of the forces that have been identified in the previous steps to work through possible futures. Create scenarios illustrating possible mitigation or redevelopment measures the community could take to safeguard these financial and community assets. The strategic options available to the community for rebuilding its community following a disaster.

Step 4: Propose Action Strategies and Steps
Action plans and steps should be developed based on various disaster scenarios, and they should be sufficiently detailed to specify the responsible stakeholder of each plan and step as well as the resources required to finish the project. All parties involved should be asked to consider these for both pre-disaster planning and post-disaster economic recovery (both short- and long-term). Once a disaster

strikes, this crucial phase will boost efficiency and assist stakeholders in making informed decisions about post-disaster recovery.
Recommendation: After a disaster, make an implementation schedule that indicates what has to be done when. This will guarantee that, in the event of a disaster, the recovery process proceeds as planned.

Step 5: Compile Contact Information
Establish a list of important contacts for the state and region's stakeholders. In the event of a disaster, the lead organization might utilize this list to coordinate information gathering and economic analysis. This should be facilitated by the establishment of a communications plan. For further in-depth details, refer to the section on Developing a Communication Strategy.

Step 6: Develop a List of Possible Funding Sources
It is crucial to identify possible federal, state, and local financing sources for disaster recovery and redevelopment.

Step 7: Follow Up with the Plan
The following actions will guarantee that the community adopts the pre-disaster economic preparedness strategy and that partnerships continue.

Step 7a: Integrate the Plan with Other Relevant Plans
In order to address discrepancies, the pre-disaster economic preparedness plan should be connected to other community plans even though it is a component of the more extensive, all-inclusive disaster planning effort. The strategy should be adopted by the local government leadership with an aim of raising awareness and guarantee execution.

Step 7b: Monitor, Evaluate, and Update the Plan

An annual review of the pre-disaster economic preparedness plan is of paramount important. The partnerships and connections that are forged among stakeholders are some of the biggest advantages of the planning process. Bringing stakeholders together once a year to review and revise the strategy is a good way to nurture these relationships (Restore Your Economy ORG, 2023). This economic preparedness plan ought to function in tandem with a more extensive, all-encompassing community planning framework headed by the local government in the event of a disaster. Coordinating it with matters like housing, infrastructure, and land use is necessary. Florida counties are currently leading the way in comprehensive planning activities connected to disasters that have a substantial economic component.

8 Conclusion

Due to the wide-ranging, global, and increasing impacts of disasters on both the economy and society, it is imperative to learn from historical events and anticipate future patterns of vulnerability. The increasing occurrence and intensity of disasters driven by climate change, urbanization, and socioeconomic inequality require that public and private sector actors work collaboratively to design and implement robust risk governance strategies. While outstanding advances have been made in modelling the physical response of infrastructure systems to disaster risks, more integrated approaches are needed to assess the broader social and economic impacts. Decision-makers must deepen their understanding of how disasters affect livelihoods, disrupt essential services, and entrench systemic inequality. This insight is vital not only for formulating effective emergency response plans but also for prioritiszing long-term mitigation investments and resilience-building strategies. Future preparedness depends on the ability to bridge scientific knowledge, policy innovation, and inclusive governance. By doing so, societies can not only reduce their exposure to disaster risks but also enhance their capacity to recover and transform in the face of future challenges.

References

Achurra-Gonzalez, P. E, Novati, M., Foulser-Piggott, R., Graham, D. J, Bowman, G., Bell, M. G. H., & Angeloudis, P. (2019). Modelling the impact of liner shipping network perturbations on container cargo routing: Southeast Asia to Europe application. *Accident Analysis & Prevention*, (123), 399–410. https://doi.org/10.1016/j.aap.2016.04.030

ADPC. (2011). *Disaster-resilient housing: A resource guide*. Asian Disaster Preparedness Center. https://www.adpc.net

Botzen, W. J. W., Deschenes, O., & Sanders, M. (2019). The economic impacts of natural disasters: A review of models and empirical studies. *Review of Environmental Economics and Policy, 13*(2), 167–188. https://doi.org/10.1093/reep/rez004

Cavallo, E., & Noy, I. (2011). Natural disasters and the economy—A survey. *International Review of Environmental and Resource Economics, 5*(1), 63–102. https://doi.org/10.1561/101.00000039

Cutter, S. L., Boruff, B. J., & Shirley, W. L. (2003). Social vulnerability to environmental hazards. *Social Science Quarterly, 84*(2), 242–261. https://doi.org/10.1111/1540-6237.8402002

European Union. (2023). *Disruption to critical infrastructure*. https://civil-protection-knowledge-network.europa.eu/eu-overview-risks/human-induced-risks/disruption-critical-infrastructure

French, S. P. (2018). Connecting physical damage to social and economic impacts. In *17th US-Japan-New Zealand workshop on the improvement of structural engineering and resilience* (pp. 4–2).

Hallegatte, S., Vogt-Schilb, A., Bangalore, M., & Rozenberg, J. (2017). *Unbreakable: Building the resilience of the poor in the face of natural disasters*. World Bank. https://openknowledge.worldbank.org/handle/10986/25335

IFRC. (2022). *What is a disaster?* International Federation of Red Cross and Red Crescent Societies. https://www.ifrc.org/our-work/disasters-climate-and-crises/what-disaster

Koks, E. E., Bockarjova, M., de Moel, H., & Aerts, J. C. J. H. (2019). Integrated direct and indirect flood risk modeling: Development and sensitivity analysis. *Risk Analysis, 35*(5), 882–900. https://doi.org/10.1111/risa.12300

Lavell, A., Oppenheimer, M., Diop, C., Hess, J., Lempert, R., Li, J., & Muir-Wood, R. (2012). *Managing the risks of extreme events and disasters to advance climate change adaptation*. Cambridge University Press. https://www.ipcc.ch/report/managing-the-risks-of-extreme-events-and-disasters-to-advance-climate-change-adaptation/

Lovell, E., & Le Masson, V. (2014). Equity and includusion in disasters risk reduction: Building resiliance for all. https://cdkn.org/sites/default/files/files/CDKN-Equity-and-inclusion-in-disaster-riskreduction-building-resilience-for-all1.pdf

Natural Institute for Communicable Deseases (NICD). (2023). Division of the national laboratory services. https://www.nhls.ac.za/wp-content/uploads/2025/03/NICD_Annual_Report_2023-2024.pdf

OECD. (2017). Socio-economic impacts of disasters in OECD countries, in Government at a Glance 2017. *OECD Publishing*. https://doi.org/10.1787/gov_glance-2017-73-en

Okuda, K., & Kawasaki, A. (2022). Effects of disaster risk reduction on socio-economic development and poverty reduction. *International Journal of Disaster Risk Reduction, 8*. https://doi.org/10.1016/j.ijdrr.2022.103241

Pelling, M., & Wisner, B. (eds). (2009). *Disaster risk reduction: Cases from urban Africa, Earthscan.*

Pregnolato, M., Ford, A., Wilkinson, S. M., & Dawson, R. J. (2017). The impact of flooding on road transport: A depth-disruption function. *Transportation Research, 55*, 67–81. https://doi.org/10.1016/j.trd.2017.06.020

Restore Your Economy. (2023). *Disaster planning for economic preparedness*. https://restoreyoureconomy.org/main/disaster-planning-for-economic-preparedness

Rodrigue, J. (2020). *The geography of transport systems*. Routledge. https://doi.org/10.4324/9780429346323

SAMHSA. (2017a). *Greater impact: How disasters affect people of low socioeconomic status*. Substance Abuse and Mental Health Services Administration. https://www.samhsa.gov

Sapountzaki, K. (2019). The interplay between socio-economic crises and disaster risks: Examples from the developed and developing World. https://www.preventionweb.net/files/65840_f213finalkalliopisapountzakitheinte.pdf?startDownload=true

Scapini, V. (2020). Disaster, infrastructure damage, and health. *International Journal of Safety and Security Engineering, 10*(2), 219–225. https://www.researchgate.net/publication/341570870

Stats SA. (2022). *Mid-year population estimates 2022*. Statistics South Africa. https://www.statssa.gov.za/

Substance Abuse and Mental Health Services Administration (SAMHSA). (2017b). *Greater impact: How disasters affect people of low socioeconomic status*. Supplemental research bulletin. U.S. Department of Health and Human Services, Office of Public Health and Science. https://www.govinfo.gov/app/details/GOVPUB-HE20-PURL-gpo106714

Tasri, E. S, Karimi, K., & Muslim, I. (2022). The effect of economic variables on natural disasters and the impact of disasters on economic variables. *Heliyon, 8*(1). https://doi.org/10.1016/j.heliyon.2021.e08678

The World Bank. (2016). *Breaking the link between extreme weather and extreme poverty*. https://www.worldbank.org/en/news/feature/2016/11/14/breaking-the-link-between-extreme-weather-and-extreme-povert

Tierney, K. (1997). Business impacts of the Northridge earthquake. *Journal of Contingencies and Crisis Management*, *5*(2), 87–97. https://doi.org/10.1111/1468-5973.00040

UNDRR. (2019). *Global assessment report on disaster risk reduction 2019*. United Nations Office for Disaster Risk Reduction. https://www.undrr.org/publication/global-assessment-report-disaster-risk-reduction-2019

Verschuur, J., Koks, E. E, & Hall, J. W. (2020). Port disruptions due to natural disasters: Insights into port and logistics resilience. *Transportation Research Part D: Transport and Environment*. https://doi.org/10.1016/j.trd.2020.102393

Wisner, B., Blaikie, P., Cannon, T., & Davis, I. (2004). *At risk: Natural hazards, people's vulnerability and disasters* (2nd ed.). Routledge.

World Health Organization (WHO). (2022). *2022 Organizational report*.

Ziervogel, G., & Smit, W. (2009). Learning to swim: Strengthening flooding governance in the city of Cape Town. Paper presented at the 2009 Amsterdam Conference on the Human Dimensions of Global Environmental Change, Amsterdam, 2–4 December 2009.

CHAPTER 7

How are Emerging Financial Instruments Transforming Sustainable Infrastructure Finance in the Face of Climate Change?

Tanishka Gupta

1 Introduction

1.1 Background: The Urgency and Scale of Sustainable Infrastructure Needs

Sustainable infrastructure is increasingly recognized as a foundational pillar in addressing the twin imperatives of climate mitigation and adaptation, while also serving as a catalyst for inclusive economic development and ecological resilience (Bhattacharya et al., 2016; Fay et al., 2019). From renewable energy grids and resilient transport systems to wastewater management and green urban ecosystems, the global demand for infrastructure that supports the Sustainable Development Goals (SDGs) and aligns with the Paris Agreement has reached unprecedented levels. Recent projections by the Global Infrastructure Facility and the OECD estimate

T. Gupta (✉)
Jagannath International Management School, New Delhi, India
e-mail: tanishka.gupta@jagannath.org

N. Tshikovhi et al. (eds.), *The BRICS Imperative*,
https://doi.org/10.1007/978-3-032-05932-1_7

that the world must invest more than USD 6.9 trillion annually through 2030 to meet sustainability-aligned infrastructure needs (Global Infrastructure Facility, 2021; OECD, 2022a). Despite this urgent demand, current flows of public sector financing and institutional capital remain grossly inadequate, resulting in a persistent and widening infrastructure financing gap—particularly acute in low- and middle-income countries (Rozenberg & Fay, 2019).

Historically, infrastructure investment has been dominated by public expenditure, bilateral development assistance, and multilateral lending. However, fiscal stress, competing social priorities, and rising sovereign debt have increasingly constrained the ability of governments to fund large-scale infrastructure projects (Helm, 2010; Humphrey, 2020). Moreover, traditional financing models have often overlooked long-term environmental and social externalities, inadequately incorporating environmental, social, and governance (ESG) metrics into project design and risk assessments (Nelson & Shrimali, 2014). This misalignment has weakened infrastructure's ability to contribute to sustainable outcomes, calling into question the viability of business-as-usual financing strategies in an era defined by climate volatility and systemic risk.

In response, policymakers and financial actors are exploring a new generation of innovative financial instruments that promise to mobilize private capital at scale, de-risk investments, and embed sustainability into the core of infrastructure finance. Instruments such as green bonds, sustainability-linked loans (SLLs), infrastructure investment trusts (InvITs), and digital-finance-based platforms like blockchain and tokenized crowdfunding are gaining traction for their potential to diversify funding sources and improve outcome accountability (Kidney & Oliver, 2014; Mazzucato & Semieniuk, 2018). These tools represent a shift from static and siloed financing models to more market-driven, performance-based, and impact-aligned frameworks. More importantly, the integration of digital technologies in sustainable finance has created new opportunities for transparency, decentralization, and stakeholder inclusion—especially relevant in contexts where public trust and financial intermediation are weak (Zetzsche et al., 2020). Yet, these innovations are not without challenges. Concerns around regulatory governance, institutional readiness, standardization, and impact verification remain critical. The efficacy of these instruments hinges not only on their financial architecture but also on the presence of supportive policy ecosystems, robust legal frameworks, and transparent reporting infrastructures (Cochu et al., 2016).

This chapter addresses the urgent need to examine how emerging financial instruments are shaping the global landscape of sustainable infrastructure finance. It analyzes their structural potential, market deployment, and governance requirements, situating them within broader dynamics of climate commitments, geopolitical uncertainty, and digital disruption. By offering a comparative and evidence-based perspective, the chapter seeks to illuminate both the possibilities and limitations of these innovations in driving a just and climate-resilient infrastructure transformation.

1.2 Problem Statement: Persistent Financing Gaps and Institutional Challenges

The paradox of our time lies not merely in the abundance of resources available in the global economy but in the manner in which these resources are mobilized—or fail to be mobilized—for public good. Nowhere is this contradiction more evident than in the domain of sustainable infrastructure financing. Despite widespread normative endorsement of sustainability goals—epitomized by the Sustainable Development Goals (SDGs) and the Paris Agreement—the financial mechanisms designed to advance such objectives remain fragmented, risk-averse, and structurally inadequate, especially for the countries of the Global South (Bhattacharya et al., 2016; UNCTAD, 2023).

The current infrastructure-financing paradigm suffers from a fundamental misalignment between economic incentives and social objectives. Multilateral commitments, though ambitious, have failed to translate into commensurate financial flows for infrastructure development, particularly in low-income and climate-vulnerable countries (Kharas & McArthur, 2021). The result is what may be termed a sustainability-finance gap, a chasm between rhetorical commitments to just and green transitions and the actual deployment of capital. This gap not only delays critical infrastructure investments but also exacerbates climatic vulnerability and economic exclusion, disproportionately affecting those already marginalized (Watkins, 2020). Traditional financing modalities—sovereign borrowing, public–private partnerships (PPPs), and commercial lending—exhibit at least three fundamental limitations. First, these models are often skewed toward short-term economic returns, with insufficient regard for long-term environmental or distributive consequences (Barlow et al., 2022). Second, they are structurally susceptible to macroeconomic shocks, such as currency volatility and political instability, which are particularly pronounced in emerging and frontier markets (Bulow & Rogoff, 2005). Third, they typically operate without embedded sustainability metrics, thereby lacking enforceable mechanisms to ensure alignment with low-carbon and socially inclusive development trajectories (Hepburn et al., 2020).

In this context, the emergence of innovative financial instruments—including green bonds, sustainability-linked loans (SLLs), infrastructure investment trusts (InvITs), and digitally enabled finance like blockchain and tokenized infrastructure assets—marks a potentially transformative shift. These instruments, in principle, aim to democratize capital flows, enhance transparency, and embed sustainability metrics into financial structuring (Zhang et al., 2022). They promise to overcome the informational asymmetries and risk aversion that have long constrained private participation in infrastructure, especially where public sector balance sheets are stretched.

Yet this optimism must be tempered with caution. Adoption remains uneven, both across geographies and sectors, reflecting deep disparities in institutional readiness and regulatory coherence (Gabor, 2021). Furthermore, pervasive concerns regarding greenwashing, the integrity of ESG disclosures, and the fragility of nascent digital governance frameworks raise critical questions about efficacy and accountability (Sharma & Ramani, 2023). Indeed, without a shared normative framework to guide financial innovation, there is a risk that these instruments may exacerbate rather than resolve existing inequities—by diverting capital toward jurisdictions with better regulatory optics but not necessarily greater need or impact (Tooze, 2021). The broader transformation of the global financial architecture—driven by decarbonization imperatives, evolving investor expectations, and interest rate volatility—further complicates the picture. While these shifts open up new opportunities for sustainability-aligned investment, they also render the system more opaque and unstable, particularly for countries and communities lacking voice in global financial governance (Gallagher & Kozul-Wright, 2022).

What is required, therefore, is not merely innovation in instruments, but innovation in governance, ethics, and institutional design. Financial tools must be embedded in frameworks that recognize both the agency of the poor and the ecological limits of growth. A justice-oriented approach to infrastructure finance must foreground not only efficiency and scalability but also equity, participation, and long-term resilience.

Therefore, the core problem this chapter seeks to address is:

> To what extent are emerging financial instruments capable of transforming the infrastructure finance ecosystem in ways that are sustainable, inclusive, and resilient?

By examining their design, implementation, and policy integration across contexts, this chapter aims to evaluate their transformative potential while identifying the systemic barriers that must be overcome to mainstream them into sustainable infrastructure finance.

2 The Evolving Landscape of Infrastructure Finance

2.1 *Traditional Infrastructure Financing Mechanisms*

Infrastructure has historically been conceived as a quintessential public good, central to the provision of essential services and the realization of collective welfare. Governments, accordingly, assumed the primary responsibility for the planning, financing, and delivery of infrastructure projects. This paradigm was characterized by strong state stewardship, underpinned by centralized planning systems, long-term public budgeting, and sovereign control over national development priorities (Estache & Fay, 2007). In developing economies, this model was augmented through official development assistance (ODA) and concessional lending from multilateral development banks, including the World Bank, Asian Development Bank, and African Development Bank, which sought to bridge both capital shortages and technical capacity gaps (Kenny, 2020).

The traditional financing architecture was structured around long project gestation periods, capital-intensive outlays, and a presumption of stable macroeconomic governance. This model enabled predictability and facilitated long-horizon planning for key development sectors such as energy access, potable water, transportation, and sanitation—areas in which market incentives alone were insufficient to ensure equitable provision (Hirschman, 1958). The national infrastructure agenda was thus predominantly funded through budgetary allocations, sovereign borrowing, and government bond markets, often underwritten by public trust in the developmental role of the state.

However, the institutional and fiscal underpinnings of this framework have come under increasing strain in recent decades. Sovereign debt ceilings, often driven by fiscal consolidation programs and macroeconomic stabilization measures, have constrained the ability of many governments to finance large-scale infrastructure through traditional means (Bova et al., 2016). Simultaneously, public–private partnerships (PPPs)—once heralded as a panacea to mobilize private capital and enhance project efficiency—have delivered uneven outcomes. Empirical evaluations suggest that PPPs have frequently suffered from imbalanced risk-sharing, opacity in contracting, and inefficiencies in sectors characterized by uncertain revenues or high regulatory complexity, such as urban sanitation and rural connectivity (Romero, 2021; Whitfield, 2012).

Crucially, these conventional instruments have failed to internalize environmental and social externalities, thereby perpetuating unsustainable development patterns. Infrastructure investments have too often proceeded with limited consideration for climate resilience, carbon intensity, or social equity, undermining the sustainability of both the physical assets and the institutional systems that support them (Hallegatte et al., 2019). This structural oversight reflects a deeper normative deficiency—namely, the absence of a framework that treats infrastructure not simply as a physical commodity or investment vehicle, but as a social entitlement and ethical imperative. Another systemic constraint arises from the mismatch between infrastructure project cash flows and investor expectations. Infrastructure assets typically exhibit long gestation periods and back-loaded revenue streams, deterring the participation of institutional investors—particularly pension funds, insurance companies, and sovereign wealth funds—that prioritize liquidity and short-to-medium term returns (Inderst, 2016). The lack of standardized de-risking frameworks, bankable project pipelines, and transparent data systems further hampers private capital mobilization (Attridge & Engen, 2019). Moreover, public and multilateral finance actors have been slow to adapt to evolving environmental, social, and governance (ESG) benchmarks, resulting in lagging alignment with global sustainability agendas.

While traditional infrastructure finance laid the foundational bedrock for national development throughout the twentieth century, its institutional design and operational modalities now appear increasingly inadequate in responding to the interlinked crises of our time—climate change, inequality, and democratic deficit. A paradigm shift is urgently needed: one that redefines finance not merely as a transactional enabler of physical capital, but as a transformative force for inclusive, resilient, and decarbonized development. Such a shift must be anchored in justice, participatory governance, and intergenerational responsibility—normative commitments that are central to any truly sustainable development framework.

2.2 *Limitations of Conventional Approaches*

While traditional infrastructure finance—centered on public expenditure, sovereign borrowing, and multilateral aid—has supported decades of global development, it now reveals serious limitations. Mounting **fiscal constraints**, **climate pressures**, and **inefficiencies in governance** have exposed the model's inability to meet the demands of the twenty-first century (Bhattacharya et al., 2016; Hallegatte et al., 2019).

As countries strive toward **SDG** and **Paris Agreement** targets, conventional approaches fall short across five critical dimensions outlined in Table 1.

Collectively, these limitations hinder infrastructure systems from becoming climate-resilient, inclusive, and financially viable. Merely increasing capital supply is insufficient; we must rethink how infrastructure is financed, governed, and aligned with broader sustainability goals.

This realization has triggered growing interest in emerging financial instruments—including green bonds, InvITs, and tokenized finance—that aim to close investment gaps, de-risk capital, and democratize access to sustainable infrastructure. The next section critically evaluates these innovations in the context of global infrastructure transformation.

2.3 *Shift Toward Sustainability and Innovation in Finance*

The constraints of traditional infrastructure finance, coupled with climate urgency, evolving investor expectations, and disruptive technological advances, are prompting a systemic reorientation of global financial systems. This transformation is driven by three major and interconnected forces: sustainability imperatives, capital market evolution, and financial innovation.

Table 1 Structural limitations of traditional infrastructure finance

Dimension	*Constraint*	*Developmental impact*	*Source(s)*
Fiscal capacity	Shrinking public budgets, rising debt service	Limits new investments and maintenance in sustainable infrastructure	Bova et al. (2016)
Sustainability misalignment	ESG metrics excluded from project appraisal	Supports carbon-intensive, inequitable projects	Hallegatte et al. (2019)
Private capital mobilization	Long gestation, poor risk profiles, weak regulation	Underutilization of institutional investor capital	Inderst and Stewart (2014); Attridge and Engen (2019)
Governance and operations	Delays, cost overruns, weak monitoring, political interference	Erodes investor trust and crowds out sustainable innovation	Flyvbjerg (2023); Romero (2021)
Climate and social risk handling	Projects ignore climate impacts and inclusivity	Reduces resilience and marginalizes vulnerable populations	Watkins (2020)

Source Literature Review

Global sustainability commitments and regulatory shifts have created pressure to align infrastructure finance with environmental and social objectives. International agreements like the Paris Agreement and the 2030 Agenda have redefined the role of infrastructure as a vehicle for sustainable development (OECD, 2020a). Regulatory frameworks such as the EU Sustainable Finance Taxonomy and TCFD guidelines are reshaping market behavior by embedding ESG compliance into fiduciary duties (EU, 2021; TCFD, 2017). Institutional investor preferences are shifting toward long-term, ESG-aligned infrastructure assets. Climate-related financial risks, the need for portfolio diversification, and growing interest in impact investing have pushed infrastructure from the periphery to the mainstream of capital allocation strategies (BlackRock, 2021; GIIN, 2022). Investor alliances such as the Net-Zero Asset Owner Alliance are amplifying pressure on project sponsors to align with science-based climate targets (UNEP FI, 2021).

Financial innovation and technology-driven tools are democratizing and decentralizing capital mobilization for infrastructure. Emerging models—including blockchain, tokenized assets, and AI-based ESG analytics—enable greater transparency, efficiency, and stakeholder inclusion (Chen et al., 2021; Zwitter et al., 2020). These tools are disrupting traditional financial intermediation and opening new pathways for retail and institutional participation alike. Together, these three forces signify a paradigmatic transition from infrastructure as a static public good to a dynamic and sustainability-oriented investment opportunity. While uneven in application, this shift marks a critical turning point in how capital is raised, allocated, and governed across infrastructure systems.

2.4 Key Trends in Global Infrastructure Investment

The evolution of infrastructure finance, much like the development trajectories it seeks to support, cannot be captured solely through the arithmetic of capital flows or the engineering of physical assets. It is shaped by an intricate interplay of normative priorities, institutional capacities, and technological possibilities (Rodrik, 2007; Stiglitz et al., 2013). What we observe today is not merely a change in instruments or actors, but a transformation in the *ethical and epistemic foundations* of how we imagine, value, and govern infrastructure (Sen, 2009). The trends outlined below reflect not only shifts in finance but shifts in purpose—revealing an emerging global consciousness that recognizes infrastructure not as a neutral artefact of economic growth, but as a central arena where questions of equity, sustainability, and resilience converge (Bhattacharya et al., 2022; UNDP, 2020). Understanding these trends, therefore, requires more than financial acumen; it demands a commitment to social reasoning, democratic participation, and a redefinition of what constitutes progress (Hickel, 2020; Sen, 1999) (Table 2).

Together, these trends point to a reconfiguration of infrastructure finance—one that blends traditional project finance expertise with emerging tools, cross-sectoral collaboration, and a shared commitment to sustainability. However, the operationalization of this shift remains uneven and fragmented, especially in developing economies where institutional and market capacities may be limited.

The next chapter turns to the emerging financial instruments that are at the forefront of this transformation, analyzing their design, function, and potential to drive sustainable infrastructure investment at scale.

Table 2 Key trends reshaping sustainable infrastructure finance in the context of climate change

Trend	*Key features*	*Primary drivers*	*Implications for infrastructure finance*
1. Climate-resilient and low-carbon infrastructure	- Focus on renewable energy (solar, wind) - Green urban infrastructure - Climate-proofing of legacy assets (roads, ports, water systems) - Nature-based solutions	- Climate policy and regulation - Rising climate risks - Investor risk awareness	- Shift from fossil-fuel assets to green projects - Need for new project assessment metrics - Greater long-term investment stability
2. Growing role of private capital and institutional investors	- Increased involvement of pension funds, insurance firms, SWFs - Blended finance and credit enhancement tools - ESG-aligned financial products	- Public finance gaps - ESG investment mandates - Risk-reward optimization	- Expansion of infrastructure as a viable asset class - More public–private partnerships (PPPs) - Mobilization of long-term "patient" capital
3. Local and subnational infrastructure finance	- Municipal green bonds - City-level PPPs - Regional financing platforms	- Urbanization and local climate exposure - Decentralization policies - Capital market access at subnational levels	- Empowerment of cities in financing - Improved resilience at the local level - Need for local capacity-building

Trend	*Key features*	*Primary drivers*	*Implications for infrastructure finance*
4. Integration of ESG and impact metrics	- Use of ESG screening tools - Impact measurement frameworks (IRIS + , GRI, SDG Impact) - Sustainability-linked bonds and loans	- Stakeholder pressure for transparency - Regulatory shifts toward mandatory disclosure - Transition from ESG compliance to impact alignment	- Enhanced accountability - ESG-performance-linked financial terms - Greater investor confidence
5. Innovation in project preparation and delivery	- Digital twins, AI, blockchain, remote monitoring - Smart procurement and lifecycle analytics - Risk modeling and pricing tools	- Need for efficiency and cost control - Advances in digital infrastructure - Data-driven decision-making	- Reduced project delays and cost overruns - More accurate risk assessment - Improved project bankability for investors

Source Literature Review

3 Emerging Financial Instruments: Typologies and Innovations

3.1 Green Bonds

Green bonds have become one of the most widely adopted financial instruments in the effort to redirect capital flows toward sustainable infrastructure. They are designed to raise funds specifically for projects with measurable environmental benefits, enabling issuers to tap into the growing market of sustainability-conscious investors. Since their launch in the late 2000s, green bonds have transitioned from a niche product to a globally recognized mechanism for aligning finance with climate and environmental goals. Green bonds are fixed-income instruments in which the proceeds are earmarked for projects that contribute to environmental sustainability, such as renewable energy, energy efficiency, sustainable transport, biodiversity conservation, and clean water systems (Flammer, 2021). The distinct feature of green bonds lies in the use-of-proceeds requirement, which mandates issuers to clearly define and report on how funds will be allocated to environmentally beneficial initiatives.

The development of international standards—such as the Green Bond Principles (ICMA, 2021) and the Climate Bonds Standard (CBI, 2022)—has provided voluntary guidelines on transparency, eligibility criteria, and post-issuance reporting. These frameworks have been instrumental in building investor trust and improving market integrity.

Since the first labeled green bond was issued by the European Investment Bank in 2007, global green bond issuance has witnessed exponential growth. According to the Climate Bonds Initiative (2023), annual issuance rose from just USD 2.6 billion in 2012 to over USD 600 billion in 2023, with cumulative issuance surpassing USD 2.6 trillion. The market is dominated by sovereigns and corporates, but development banks, municipalities, and financial institutions are also key players. Academic studies (Karpf & Mandel, 2018; Zerbib, 2019) confirm that green bonds often enjoy strong investor demand and, in some markets, benefit from a "greenium"—a slight yield advantage reflecting high ESG demand.

3.1.1 European Investment Bank's Green Bond Initiative

The European Investment Bank (EIB) issued the world's first labeled green bond in 2007, termed the "Climate Awareness Bond" (CAB).

This instrument marked the beginning of a dedicated effort to finance projects aligned with climate mitigation goals, particularly in the fields of renewable energy and energy efficiency. The EIB's issuance was groundbreaking in that it coupled a transparent use-of-proceeds model with detailed impact reporting—practices that are now standard in the green bond market (EIB, 2022) (Fig. 1).

A review by Park (2018) notes that the EIB's green bond model significantly influenced the development of the Green Bond Principles and other sovereign green bond frameworks. The initiative also paved the way for the EU Green Bond Standard, which seeks to codify best practices, reduce fragmentation in the green finance market, and curb greenwashing. The EIB has issued over EUR 50 billion in green bonds across global markets, positioning itself as both a financial intermediary and a norm-setting institution. Its experience illustrates the potential of

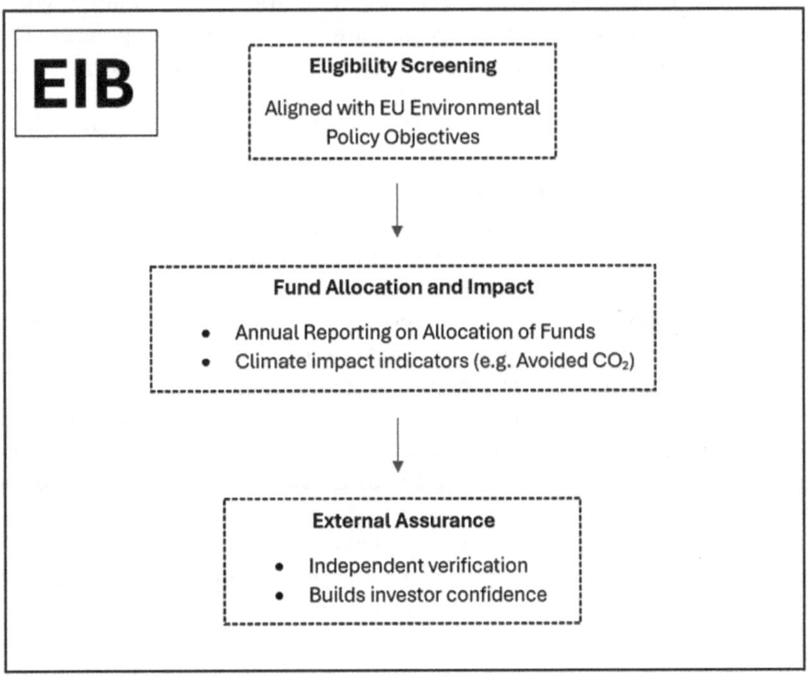

Fig. 1 Key features of EIB model; (*Source* Author)

development banks to crowd in private capital through demonstrative leadership in financial innovation.

Green bonds have proven effective in attracting a new class of investors focused on environmental impact, thereby deepening capital markets for infrastructure (OECD, 2021). The expansion of green bond indices and ESG mutual funds has increased liquidity and diversified sources of demand. The use-of-proceeds requirement, combined with third-party certification and post-issuance reporting, enhances transparency and improves project governance (Tang & Zhang, 2020). Empirical studies (e.g., Ehlers & Packer, 2017; Zerbib, 2019) provide evidence of a "greenium" a modest but consistent yield discount that issuers enjoy due to strong investor demand for green assets. Green bonds have encouraged regulatory innovation and have often been the entry point for countries developing green taxonomies or launching sovereign ESG frameworks (World Bank, 2021).

One of the major risks is the potential for greenwashing, where proceeds are allocated to projects that are only nominally "green." The absence of globally binding definitions exacerbates this problem (Schinasi et al., 2022). Despite their global growth, green bonds are largely concentrated in upper-middle and high-income countries. In low-income contexts, limited creditworthiness, shallow capital markets, and lack of technical capacity hinder adoption (UNEP FI, 2023). Issuing green bonds often involves additional costs related to verification, certification, and monitoring—barriers that deter smaller issuers or first-time participants (CBI, 2022). The requirement that proceeds be used only for predefined green projects can be restrictive, especially when infrastructure needs are cross-cutting or when broader social infrastructure lacks classification under green finance frameworks (Banga, 2019).

Green bonds have played a pivotal role in mainstreaming environmental considerations into infrastructure finance. They demonstrate how market-based instruments, coupled with transparency and accountability mechanisms, can mobilize large-scale capital for climate-aligned infrastructure. However, their future impact will depend on the evolution of global standards, the integrity of impact verification, and the ability to extend access to markets that need sustainable infrastructure the most.

3.2 *Sustainability-Linked Loans (SLLs)*

While green bonds earmark proceeds for specific environmentally beneficial projects, Sustainability-Linked Loans (SLLs) represent a more flexible and performance-driven approach to sustainable finance. These instruments are not restricted by the use-of-proceeds principle; instead, their key innovation lies in linking the loan terms—typically the interest rate—to the borrower's achievement of predefined sustainability performance targets (SPTs). This makes SLLs particularly appealing for corporates and infrastructure developers whose sustainability goals span across operations, asset classes, or geographies.

SLLs are structured as conventional term loans or revolving credit facilities, but include sustainability-linked provisions that modify the cost of capital based on the borrower's performance against agreed ESG metrics. These provisions are guided by the Sustainability-Linked Loan Principles (SLLP) issued by the Loan Market Association (LMA), Asia Pacific Loan Market Association (APLMA), and Loan Syndications and Trading Association (LSTA) (LMA, 2023).

Sustainability Performance Targets (SPTs): These are quantifiable and time-bound ESG-related goals. Examples include reduction in greenhouse gas (GHG) emissions, improved energy efficiency, increased share of renewables in energy mix, or enhanced water recycling rates.

SPTs are measured using KPIs that must be material, externally verifiable, and benchmarked against industry standards or third-party frameworks (e.g., SBTi, GRI). The interest margin on the loan is adjusted usually downward if the borrower meets or exceeds the agreed SPTs. Conversely, failing to meet targets can lead to upward pricing adjustments or reputational consequences. Regular third-party verification (often annually) is required to ensure the credibility of performance reporting and to avoid sustainability-washing. SLLs are particularly suited to entities pursuing enterprise-wide sustainability transitions, as they allow capital to be used flexibly while holding borrowers accountable for outcomes. As such, they appeal to both corporates with ESG roadmaps and infrastructure developers with decarbonization or resilience targets.

Recent empirical studies suggest that SLLs can improve firms' environmental performance while also reducing their cost of debt over time. Moreover, SLLs help mainstream sustainability goals into financial decision-making processes, moving ESG considerations from peripheral reporting to core financial incentives. Although SLLs have gained initial traction in corporate and industrial sectors (e.g., manufacturing, real estate), they are increasingly being applied in infrastructure finance, particularly where ESG outcomes are clearly measurable and project-based financing is less viable. Notable use cases include:

Energy companies are using SLLs to support transitions from fossil fuels to renewable sources. For instance, Enel, the Italian utility, issued a €2.5 billion SLL linked to its target of increasing its renewable energy capacity to 55% by 2025. This facility enabled Enel to fund a range of activities including grid modernization and solar installations while being rewarded with lower interest rates upon achieving renewable benchmarks (ICMA, 2022). Infrastructure firms managing urban water and sanitation systems have utilized SLLs to incentivize reductions in non-revenue water, improvements in wastewater recycling, and energy efficiency in treatment plants. In 2020, Manila Water in the Philippines secured an SLL with KPIs related to water loss reduction and service coverage in underserved communities, tying financial benefits directly to sustainability and access outcomes (ADB, 2021). Logistics companies and urban transit operators are exploring SLLs to finance electrification of fleets, modal shifts from road to rail, and implementation of low-emission zones. These loans allow for flexibility in applying funds across multiple projects while maintaining a clear performance commitment on emissions or modal share. Some borrowers combine environmental and social KPIs, such as gender inclusion in infrastructure labor forces, occupational safety improvements, or affordable housing targets. The hybrid nature of SLLs makes them suitable for blended ESG strategies, particularly in integrated infrastructure development.

Despite their versatility, the widespread adoption of SLLs in infrastructure remains nascent. Challenges includes—Defining sector-specific KPIs that are both ambitious and measurable; Ensuring regulatory alignment, particularly in public–private projects; Managing data availability and verification costs, which may be prohibitive for smaller developers or municipalities. Nonetheless, the trend is accelerating. According to BloombergNEF (2023), global SLL issuance reached USD 700 billion by the end of 2022, with infrastructure-linked deals accounting for an increasing share—particularly in Europe and Asia–Pacific. As frameworks mature and ESG data becomes more standardized, SLLs are expected to become an integral part of infrastructure finance portfolios, offering a bridge between corporate finance flexibility and infrastructure sector impact accountability.

3.3 *Infrastructure Investment Trusts (InvITs)*

Infrastructure Investment Trusts (InvITs) represent an institutional innovation aimed at reconfiguring the financial architecture of infrastructure development, particularly in contexts where fiscal constraints and underdeveloped capital markets hinder long-term investment (Bhattacharya et al., 2016; Ghosh & Ganesan, 2021). Emerging at the intersection of public infrastructure needs and private capital imperatives, InvITs provide a platform for asset monetization while broadening the base of investors who can participate in infrastructure finance. Their significance lies not merely in mobilizing resources but in altering the modality of participation—shifting from closed, illiquid models of project finance to more transparent, yield-linked mechanisms that align infrastructure with financial market logic (Inderst & Stewart, 2014). Functionally, InvITs operate as collective investment vehicles, drawing conceptual lineage from Real Estate Investment Trusts (REITs), yet adapted to the long-term, operational characteristics of infrastructure assets such as highways, transmission networks, and renewable energy installations. These vehicles allow infrastructure developers to divest mature, revenue-generating assets, thereby recycling capital into new projects without compromising on returns or operational integrity (KPMG, 2020). For investors—both institutional and retail—InvITs offer access to stable cash flows, diversified risk, and regulated transparency under the oversight of securities authorities such as the Securities and Exchange Board of India (SEBI).

The adoption of InvITs marks a paradigmatic shift: infrastructure is no longer seen solely as a public good managed by state authorities but increasingly as a financial asset class capable of delivering social utility and economic returns in tandem. This reimagination, while promising, also raises deeper questions about accessibility, equity, and the public interest—especially in settings where the democratization of finance is still nascent and regulatory capacities remain uneven.

InvITs operate under two primary models:

- Publicly Listed InvITs

These are listed on stock exchanges and open to both institutional and retail investors. Units are tradable, offering liquidity similar to equity markets. This model is suitable for large-scale, stabilized assets with predictable cash flows.

- Private Placement InvITs

These are offered to a select group of institutional investors such as pension funds, sovereign wealth funds, and insurance companies. They are not listed, offering more flexibility but lower liquidity. This model is often preferred for more complex or early-stage infrastructure portfolios (Table 3).

Investors earn returns from the net distributable cash flows generated by the underlying assets. Projects included are typically operational and revenue-generating, reducing project development risk. InvITs are governed by detailed disclosure norms, valuation standards, leverage caps, and minimum distribution obligations (e.g., 90% of cash flow to be distributed in India, per SEBI norms). Infrastructure developers can free up capital locked in completed assets and reinvest in new projects, enhancing capital efficiency. Scholarly reviews (e.g., Singh, 2021) indicate that InvITs can reduce financing costs, lengthen investment horizons, and foster institutional confidence in emerging market infrastructure. However, their uptake depends heavily on robust governance frameworks, tax clarity, and investor awareness.

3.3.1 *India's InvIT Experience in Roads and Energy*

India has emerged as a frontrunner in the adoption of Infrastructure Investment Trusts (InvITs), establishing itself as a key innovator in infrastructure finance among Global South economies. Confronted by the twin challenges of expanding infrastructure needs and mounting fiscal constraints, the Government of India instituted a regulatory framework for InvITs in 2014 through the Securities and Exchange Board of India (SEBI)—a pioneering move for the region (Ghosh & Ganesan, 2021; SEBI, 2014). Since their formalization, InvITs have gained significant

Table 3 InvIT models in India

Name	*Description*	*Investor type*	*Liquidity*	*Asset suitability*	*Regulatory/ listing status*
Publicly listed InvITs	Listed on stock exchanges and accessible to both institutional and retail investors. Units are tradable. (PwC India, 2020)	Institutional and retail investors	High (due to stock market listing)	Large-scale, stabilized assets with predictable cash flows (ICRA, 2022)	SEBI-regulated; listed on stock exchanges (SEBI, 2014)
Private place-ment InvITs	Offered to select institutional investors; not listed, offering flexibility but lower liquidity. (KPMG, 2021)	Institutional only (e.g., pension, SWFs, insurers) (NIIF, 2023)	Low (not publicly traded) (KPMG, 2021)	Complex or early-stage infrastructure portfolios (PwC India, 2020)	SEBI-regulated; unlisted private offering (SEBI, 2014)

Source Literature Review

traction, particularly in sectors where revenue predictability and asset maturity enhance their attractiveness to long-term investors, such as roads and power transmission (World Bank, 2022a).

The National Highways Authority of India (NHAI) demonstrated the public-sector application of InvITs in 2021 by monetizing operational toll roads under the Bharatmala Pariyojana programme. The initial tranche, encompassing 390 kilometers across five highway stretches, raised over ₹5,000 crore (approximately USD 600 million), attracting capital from marquee institutional investors including the Canada Pension Plan Investment Board (CPPIB) and the Ontario Teachers' Pension Plan (MoRTH, 2021). Under a concession-based structure, toll collection rights were transferred to the InvIT, with proceeds reinvested into new greenfield road infrastructure via public–private partnerships (PPPs). This shift marked a paradigmatic movement from direct public provisioning toward a market-mediated approach, deepening capital market participation in national infrastructure financing (ADB, 2023).

In parallel, PowerGrid Corporation of India—the country's central transmission utility—launched its InvIT to monetize brownfield transmission assets in 2021. The initial public offer raised ₹7,735 crore (approximately USD 1.03 billion), with strong domestic and foreign investor interest. The assets, spanning multiple Indian states, were supported by long-term tariff certainty via regulatory contracts, making them ideal candidates for the InvIT model (PowerGrid, 2021). Inspired by PowerGrid's success, other energy entities, particularly in renewable power generation, are now exploring similar instruments—underscoring how predictable power purchase agreements (PPAs) enhance financial structuring in wind and solar infrastructure. InvITs offer a dual advantage: they enable sponsors to monetize de-risked, revenue-generating assets (brownfields) while recycling capital into new development (greenfields). Simultaneously, they offer low-risk entry points for institutional investors seeking stable, yield-generating exposure to infrastructure (Inderst & Stewart, 2014). Publicly listed InvITs further support liquidity, standardized disclosure, and corporate governance practices that mirror those of listed equity markets, improving investor confidence.

Certain structural challenges persist. InvIT adoption remains concentrated in roads and power, with limited diversification into social or urban infrastructure. Liquidity constraints may arise for privately placed InvITs, especially under macroeconomic stress. Although InvITs in India enjoy tax pass-through status, ambiguities in cross-border tax treatment remain a deterrent for some foreign investors (World Bank, 2022b). Further, the development of active secondary markets and standardized valuation methodologies is still evolving, posing risks to price discovery and asset transparency. Institutional limitations among InvIT sponsors and trustees may further erode effective project oversight and accountability.

As institutional mechanisms, InvITs reflect a deeper transformation in the role of finance—from a transactional tool of fiscal relief to a strategic enabler of sustainability-oriented infrastructure development. While InvITs do not inherently embed environmental or social safeguards, their structural design enables alignment with climate goals and inclusive development—particularly when paired with robust ESG integration frameworks. In the context of developing economies like India, where the capacity to deliver large-scale infrastructure is often compromised by budgetary limitations, InvITs provide a credible model for crowding in private capital while reducing the sovereign debt burden and ensuring long-term service delivery.

3.4 Fintech and Blockchain-Enabled Models

The convergence of financial technology (fintech) and blockchain is unlocking new frontiers in infrastructure finance—particularly in emerging markets where conventional financing mechanisms are constrained by high risk, low transparency, and limited institutional reach. Fintech platforms are enabling broader participation, reducing transaction costs, and improving monitoring in infrastructure projects, while blockchain technologies are being explored to ensure transparency, enhance trust, and streamline the flow of capital in long-gestation projects. These innovations hold significant potential to democratize finance, facilitate retail and diaspora investment, and address governance challenges that have traditionally plagued infrastructure delivery, especially in the Global South.

Crowdfunding—once confined to creative industries and startups—is being adapted for small- and medium-scale infrastructure, particularly renewable energy, community water systems, and urban microgrids. Using digital platforms, investors (including retail and diaspora investors) can contribute directly to projects in return for fixed returns, revenue shares, or project tokens. Platforms such as TRINE (Sweden) and WeShareSolar (USA) have successfully financed off-grid solar infrastructure in Africa and South Asia using this model. Crowdfunding platforms benefit from—Lower transaction and intermediation costs, Global investor outreach, and Real-time investment tracking and updates. The regulatory ambiguity, risk profiling, and scale limitations still pose challenges to mainstreaming infrastructure crowdfunding in national finance ecosystems (UNDP, 2022).

Tokenization refers to the digitization of real-world assets—such as toll roads, power plants, or public buildings—into tradeable blockchain-based units (tokens). This model allows fractional ownership, enabling smaller investors to access infrastructure assets typically reserved for institutional capital. Liquidity and secondary trading, where tokens can be bought and sold like securities; Transparency and traceability, since blockchain records ownership history; Lower capital entry thresholds, enhancing financial inclusion. A study by OECD (2023) notes that tokenized infrastructure may enable "financial deepening" in frontier markets, where conventional capital markets are underdeveloped. However, concerns around custodianship, legal enforceability, and smart contract risk remain barriers. Smart contracts are self-executing digital contracts that run on blockchain protocols. They can automate disbursements (e.g., releasing funds upon completion of project milestones), enforce performance-based payments, and reduce corruption or delay in fund allocation.

In infrastructure contexts, smart contracts can ensure timely contractor payments, Link financial incentives to performance indicators (e.g., emissions reduction), Support real-time monitoring of project compliance. Proof-of-concept pilots by the World Bank and African Development Bank (AfDB) have demonstrated how blockchain-led disbursement models could reduce leakages in large-scale public infrastructure projects (World Bank, 2022c).

3.4.1 *Blockchain and Transparency in African Infrastructure Projects*

A compelling application of blockchain for infrastructure governance has emerged in Africa, where transparency and accountability are critical yet often weak due to institutional bottlenecks and corruption risks. In response, several pilots and platforms have begun exploring blockchain-based tracking of public infrastructure investments to build trust with stakeholders and donors.

In 2020, Ghana's Ministry of Finance—in collaboration with UNDP and private sector partners—launched a blockchain pilot to track disbursement and procurement processes for school and road construction projects under its national infrastructure program. This plan included notable key feature like immutable tracking of fund flows from national treasury to implementing agencies; Digitized procurement records, reducing the chance of tampering; Community-level monitoring via mobile apps, where citizens could report discrepancies in real time.

Preliminary findings showed improved procurement transparency, reduced delays in disbursements, and greater civil society trust in project outcomes (UNDP Ghana, 2021). Immutable records reduce fraud and rent-seeking in large infrastructure contracts. Tokenization and crowdfunding enable broader participation beyond institutional capital. Smart contracts automate compliance and payments, reducing administrative overheads. Enhanced visibility into project performance supports adaptive governance. Most jurisdictions lack clear legal frameworks for blockchain-based finance. Smart contract bugs, cyberattacks, and platform failures pose new systemic risks. Most applications remain pilots or localized experiments, not yet viable for national-scale infrastructure pipelines. Low internet penetration or technical literacy in some regions could hinder adoption. Fintech and blockchain-based innovations represent a paradigm shift in how infrastructure finance can be democratized, governed, and delivered. While these models are not substitutes for mainstream instruments like green bonds or InvITs, they serve as critical complements—particularly in fragile contexts, decentralized systems, and innovative ESG-linked ecosystems. Their maturation will require interoperable regulations, digital public infrastructure, and sustained cross-sectoral experimentation.

4 How Finance Supports Environmental Goal Achievement

Green Bonds have made significant strides in aligning financial markets with climate goals. According to the Climate Bonds Initiative (2023), global green bond issuance reached over USD 600 billion in 2023. Bonds issued by the European Investment Bank and sovereigns like France, India, and Nigeria have been used to fund renewable energy, clean transport, and energy efficiency projects. For example, Nigeria's sovereign green bond (2017) financed off-grid solar systems and reforestation efforts (World Bank, 2021).

Sustainability-linked loans (SLLs) further support climate goals by embedding performance-based incentives into financial contracts. Companies like Enel (Italy) and Manila Water (Philippines) have tied loan terms to carbon reduction and water efficiency, respectively, integrating ESG metrics into core financing (Table 4).

4.1 Resilience: Long-Term Risk Mitigation and Adaptability

In reimagining the financial architecture for sustainable infrastructure, it is imperative to look not merely at the availability of capital but at the instruments through which that capital is mobilized, governed, and translated into resilience. The following table maps how emerging financial tools operate across dimensions of risk mitigation, adaptive capacity, and implementation complexity—revealing that their transformative potential lies not in novelty alone, but in the integrity, inclusiveness, and insight with which they are deployed (Table 5).

Table 4 Emerging financial instruments for sustainable infrastructure and climate impact

Instrument	*Environmental focus*	*Examples*	*Impact metrics*	*References*
Green bonds	Project-based funding for environmentally beneficial projects	European Investment Bank (EIB), Nigeria Sovereign Green Bond, Indian Railways Green Bond	Emissions avoided, renewable capacity added, climate adaptation benefits	EIB (2023); ICMA (2021); World Bank (2022d)
Sustainability-linked loans (SLLs)	Corporate financing linked to sustainability performance targets	Enel (GHG emission targets), Manila Water (leakage reduction KPIs)	KPI compliance, ESG target performance, margin step-ups/down	LMA (2023); OECD (2022b)
Infrastructure investment Trusts (InvITs)	Indirect impact—depends on the nature of underlying infrastructure assets	PowerGrid InvIT (low-carbon transmission), IRB InvIT	Grid efficiency, emissions intensity, capital recycling	SEBI (2014); PwC India (2020); NIIF (2023)
Blockchain-enabled models	Enables transparent, programmable, and traceable financing for climate infrastructure	Ghana's Ministry of Finance road fund tracking; solar tokenization pilots	Real-time audits, traceable disbursement, fraud reduction	World Economic Forum (2022); UNDP (2021)

Source Literature Review

Table 5 Evaluating climate finance instruments through risk mitigation, resilience, and limitations

Instrument	*Risk mitigation mechanism*	*Resilience enhancement*	*Limitation/challenge*	*Interpretation & strategic insight*
Green bonds	Sovereign or Multilateral Development Bank (MDB) guarantees; third-party verification of proceeds	Funds climate-resilient infrastructure directly (e.g., flood-proof transport, renewables)	Greenwashing risk; rigid use-of-proceeds restricts flexibility	Effective for financing clearly defined green projects, but needs strong frameworks to ensure impact authenticity and adaptability
SLLs (sustainability-linked loans)	Interest rate adjustments based on sustainability performance (KPI-based repricing)	Incentivizes firms to meet climate adaptation or ESG targets over time	Relies on robust monitoring, data verification, and credible KPIs	Encourages behavioral change in borrowers, but only effective when backed by transparent governance and ESG integrity

(continued)

Table 5 (continued)

Instrument	*Risk mitigation mechanism*	*Resilience enhancement*	*Limitation/challenge*	*Interpretation & strategic insight*
InvITs (infrastructure investment trusts)	Asset diversification, long-term cash flow predictability, and pooling reduce volatility	Channels institutional capital into climate-resilient infrastructure with lower perceived risk	Asset portfolios may lack green focus or sectoral diversity	Suitable for stable, large-scale assets; requires clearer ESG screening or greening standards to maximize climate alignment
Blockchain-based models	Tamper-proof smart contracts, decentralized audit trails ensure transparency and reduce fraud	Enhances trust in governance and allocation of climate finance; enables real-time oversight	Technological scalability, regulatory acceptance, and digital infrastructure gaps	Powerful for ensuring trust and traceability in infrastructure finance; best used in combination with traditional instruments in phased rollouts

Source Literature Review

5 Conclusion

Emerging financial instruments—such as green bonds, sustainability-linked loans (SLLs), infrastructure investment trusts (InvITs), and fintech- or blockchain-enabled models—are beginning to reshape the global infrastructure finance architecture. These tools provide fresh, potentially transformational pathways for aligning infrastructure development with pressing climate objectives, social equity imperatives, and long-term economic resilience (Boffo & Patalano, 2020; Flammer, 2021).

The findings in this chapter underscore that while no single instrument is a panacea, each contributes uniquely to the broader evolution of sustainable infrastructure finance. Green bonds serve as project-specific vehicles for channeling climate-focused capital into areas like renewable energy, sustainable transport, and flood-resilient infrastructure (EIB, 2023). SLLs, by contrast, embed ESG performance targets directly into loan terms, providing dynamic incentives for improved environmental and social outcomes (LMA, 2023). InvITs facilitate asset monetization and democratize infrastructure ownership through public trust structures that appeal to both institutional and retail investors (PwC India, 2020). Meanwhile, fintech and blockchain-based models offer transparency, auditability, and decentralized access to climate finance—especially valuable in contexts with low trust in public systems or limited intermediary capacity (Arner et al., 2022; UNDP, 2021).

Despite their promise, the transformative potential of these instruments is not automatic. It hinges on the presence of enabling conditions such as robust capital markets, standardized ESG taxonomies, transparent data ecosystems, and institutional maturity (Ameli et al., 2021). In many emerging and low-income contexts, constraints such as regulatory lag, fragmented markets, and limited scalability of fintech pilots continue to inhibit broader adoption (Brueckner et al., 2022; Le & Phan, 2023).

5.1 Policy Recommendations

To fully harness the promise of these financial innovations for sustainable, inclusive, and resilient infrastructure development, the following multi-level policy actions are proposed (Table 6).

Table 6 Policy recommendations

Policy action	*Key focus*	*Strategic measures*	*Examples/references*
1. Develop robust and harmonized ESG taxonomies	Prevent greenwashing and enhance cross-border investment confidence	- Create interoperable ESG classifications across jurisdictions - Use existing standards (e.g., EU Green Bond Standard) as templates	ASEAN Green Bond Standards (OECD, 2020b); Brueckner et al. (2022)
2. Scale blended finance and credit enhancement	De-risk investments in developing economies	- Mobilize concessional finance - Deploy guarantees and technical assistance	GIF, IFC Co-Lending Program (OECD & WEF, 2021)
3. Invest in fintech infrastructure and legal frameworks	Enable scaling of blockchain and tokenized models	- Develop digital ID, e-KYC, and smart contract legality - Support sandbox environments for innovation	Ghana blockchain pilot (UNDP, 2021)
4. Mandate ESG data collection and disclosure	Ensure integrity of performance-linked finance	- Standardize ESG metrics - Mandate disclosure and ESG ratings - Support small issuers with certification subsidies	India's BRSR framework (Boffo & Patalano, 2020)
5. Decentralize access to finance	Address spatial inequalities and local infrastructure needs	- Support local issuance of green bonds - Enable tokenized municipal finance and crowdfunding	Curitiba municipal bonds (NIIF, 2023)
6. Empower ecosystem builders and anchor institutions	Build institutional capacity for financial innovation	- Fund and mandate green banks, innovation labs, and investment platforms	UK Green Investment Bank; India's NIIF (Inderst & Stewart, 2014)
7. Promote South-South cooperation and policy learning	Foster mutual learning among developing economies	- Institutionalize peer platforms (e.g., IBSA, BRICS) - Share models for InvITs, fintech governance, ESG regulation	

Source Chapter Findings/Literature Review

5.2 *Limitations*

Emerging financial instruments—such as green bonds, sustainability-linked loans (SLLs), InvITs, and fintech-based models—are redefining infrastructure finance by enabling climate-aligned investment. Yet their impact is uneven, with developed economies better positioned to leverage these tools due to mature markets, regulatory stability, and institutional readiness (Ameli et al., 2021; Inderst & Stewart, 2014). In contrast, emerging economies face structural constraints including shallow capital markets, limited ESG data, and weak governance. Many instruments, particularly green bonds and InvITs, remain largely exclusive to institutional investors, curbing inclusivity and public participation (Flammer, 2021). While digital models like crowdfunding and blockchain show promise for democratizing access, they are hindered by regulatory uncertainty and cybersecurity risks (Le & Phan, 2023). Broader adoption is further constrained by the absence of standardized taxonomies, illiquid secondary markets, and inconsistent ESG metrics—factors that erode transparency and investor trust (Boffo & Patalano, 2020; Brueckner et al., 2022; OECD, 2020). Regulatory inertia around fintech and tokenized assets adds to the challenge (Arner et al., 2022). Overcoming these barriers will require harmonized disclosure norms, inclusive investment structures, and enhanced institutional capacity—particularly in the Global South—to prevent deepening disparities in access to sustainable infrastructure finance.

References

ADB (2021). *Sustainability-Linked loans for infrastructure in Asia.*

ADB (2023). *Developing capital markets for infrastructure finance: Insights from India's InvIT experience.* Asian Development Bank.

AfDB (2021). *Smart contracts and public infrastructure.*

Ameli, N., Drummond, P., Bisaro, A., Grubb, M., & Chenet, H. (2021). Climate finance and disclosure for institutional investors: Why transparency is not enough. *Climatic Change, 167*(3–4), 1–24. https://doi.org/10.1007/s10584-021-03117-2

Arner, D. W., Barberis, J. N., & Buckley, R. P. (2022). FinTech, RegTech and the reconceptualization of financial regulation. *Northwestern Journal of International Law & Business, 41*(2), 273–312.

Attridge, S., & Engen, L. (2019). *Blended finance in the poorest countries: The need for a better approach*. Overseas Development Institute (ODI). https://odi.org/en/publications/blended-finance-in-the-poorest-countries/

Banga, J. (2019). The green bond market: a potential source of climate finance for developing countries. *Journal of Sustainable Finance & Investment, 9*(1).

Barlow, M., Clarke, T., & O'Callaghan, J. (2022). Beyond PPPs: Public investment, climate justice, and the future of infrastructure. *Development, 65*(1), 20–28.

Bhattacharya, A., Kharas, H., Plant, M., & Prizzon, A. (2022). *Financing sustainable infrastructure in the decade of action: A global agenda*. Brookings Institution.

Bhattacharya, A., Oppenheim, J., & Stern, N. (2016). *Driving sustainable development through better infrastructure: Key elements of a transformation program*. Brookings Institution. https://www.brookings.edu/research/driving-sustainable-development-through-better-infrastructure/

BlackRock. (2021). *Sustainability as BlackRock's new standard for investing*. https://www.blackrock.com/corporate/about-us/sustainability-resilience

BloombergNEF (2023). Sustainable Debt Market Outlook.

Boffo, R., & Patalano, R. (2020). *ESG investing: Practices*. OECD. https://doi.org/10.1787/734f0eb7-en

Bova, E., Ruiz-Arranz, M., Toscani, F., & Ture, E. (2016). *The fiscal costs of contingent liabilities: A new dataset*. IMF Working Paper No. 16/14. https://www.imf.org/en/Publications/WP/Issues/2016/12/31/The-Fiscal-Costs-of-Contingent-Liabilities-A-New-Dataset-43677

Brueckner, M., Comtois, A., & Westermann, A. (2022). Greenwashing in bond markets: Evidence from the European Union. *Journal of Sustainable Finance & Investment, 12*(2), 213–233. https://doi.org/10.1080/20430795.2021.1913125

Bulow, J., & Rogoff, K. (2005). Grants versus loans for development banks. *American Economic Review, 95*(2), 393–397. https://doi.org/10.1257/000282805774669879

Chen, C., Zhang, W., & Wu, L. (2021). Blockchain-based financing models for infrastructure projects: Opportunities and risks. *Technological Forecasting and Social Change, 173*, Article 121113.

Climate Bonds Initiative (2023). *Green bond market summary*.

Cochu, A., Glenting, C., Hogg, D., Georgiev, I., & Afanasyeva, E. (2016). *Study on the potential of green bond finance for resource-efficient investments*. European Commission. https://ec.europa.eu/environment/enveco/pdf/potential-green-bond.pdf

Ehlers, T., & Packer, F. (2017). Green bond finance and certification. *BIS Quarterly Review*.

EIB. (2023). *Sustainability Report 2023*. European Investment Bank. https://www.eib.org

Estache, A., & Fay, M. (2007). *Current debates on infrastructure policy*. World Bank Policy Research Working Paper No. 4410. https://openknowledge.worldbank.org/handle/10986/7370

EU. (2021). *EU Taxonomy Climate Delegated Act*. European Commission.

European Investment Bank (2022). *Sustainability report*.

European Investment Bank (EIB). (2023). *Sustainability report 2023*. https://www.eib.org

Fay, M., Hallegatte, S., Vogt-Schilb, A., Rozenberg, J., Narloch, U., & Hee, B. (2019). *Infrastructure for development: A policy agenda*. World Bank. https://doi.org/10.1596/978-1-4648-1322-9

Flammer, C. (2021). Corporate green bonds. *Journal of Financial Economics, 142*(2), 499–516. https://doi.org/10.1016/j.jfineco.2021.05.014

Flyvbjerg, B. (2023). *How big things get done*. Currency.

Gabor, D. (2021). The wall street consensus. *Development and Change, 52*(3), 429–459. https://doi.org/10.1111/dech.12645

Gallagher, K. P., & Kozul-Wright, R. (2022). *The case for a New Bretton woods: Rebuilding the global financial architecture for sustainable development*. UNCTAD Research Paper No. 80. https://unctad.org/system/files/official-document/ser-rp-2022d2_en.pdf

Ghana Ministry of Finance (2021). *Blockchain infrastructure pilot report*.

Ghosh, S., & Ganesan, K. (2021). Infrastructure investment trusts: Unlocking capital and efficiency. *Energy Policy Journal, 155*, Article 112308.

GIIN. (2022). *Annual impact investor survey*. Global Impact Investing Network.

Global Infrastructure Facility. (2021). *Annual report 2021: Catalyzing private capital for infrastructure*. https://www.globalinfrafacility.org

Hallegatte, S., Rentschler, J., & Rozenberg, J. (2019). *Lifelines: The resilient infrastructure opportunity*. World Bank. https://doi.org/10.1596/978-1-4648-1430-1

Helm, D. (2010). Infrastructure and infrastructure finance: The role of the state. *Oxford Review of Economic Policy, 26*(3), 248–264. https://doi.org/10.1093/oxrep/grq017

Hepburn, C., O'Callaghan, B., Stern, N., Stiglitz, J. E., & Zenghelis, D. (2020). Will COVID-19 fiscal recovery packages accelerate or retard progress on climate change?*Oxford Review of Economic Policy, 36*(Supplement_1), S359–S381. https://doi.org/10.1093/oxrep/graa015

Hickel, J. (2020). *Less is more: How degrowth will save the world*. Penguin.

Hirschman, A. O. (1958). *The strategy of economic development*. Yale University Press.

Humphrey, C. (2020). *Time for a new approach to funding infrastructure in developing countries*. ODI Working Paper. https://cdn.odi.org/media/documents/200611_new_approach_infrastructure_wp_final.pdf

ICMA (2021). *Green bond principles: Voluntary process guidelines for issuing green bonds*. International Capital Market Association.

ICMA (2022). *Sustainability-Linked bond and loan principles—Case studies*. International Capital Market Association.

ICRA. (2022). *Indian InvITs: Growing acceptance among investors; Asset Monetization Picking Up Pace*. https://www.icra.in

Inderst, G. (2016). *Infrastructure investment, private finance, and institutional investors: Asia from a global perspective. Asian Development Bank Institute Working Paper* No. 555. https://www.adb.org/publications/infrastructure-investment-private-finance-and-institutional-investors

Inderst, G., & Stewart, F. (2014). *Institutional investment in infrastructure in developing countries: Introduction to potential models*. World Bank Policy Research Working Paper No. 6780. https://doi.org/10.1596/1813-9450-6780

Karpf, A., & Mandel, A. (2018). *The changing value of the 'green' label on the US municipal bond market*. Nature Climate Change.

Kenny, C. (2020). *Infrastructure investment and development: Who benefits and who pays? Center for Global Development Policy Paper* 170. https://www.cgdev.org/publication/infrastructure-investment-and-development-who-benefits-and-who-pays

Kharas, H., & McArthur, J. W. (2021). *Building the SDG economy: Needs, spending, and financing for universal achievement of the sustainable development goals*. Brookings Global Working Paper Series. https://www.brookings.edu/research/building-the-sdg-economy/

Kidney, S., & Oliver, P. (2014). *Creating green bond markets: Insights, innovations, and tools from emerging markets*. International Finance Corporation & Climate Bonds Initiative.

KPMG. (2020). *InvITs—Revolutionizing infrastructure financing in India*. KPMG India.

KPMG. (2021). *Infrastructure Investment Trusts (InvITs) and Real Estate Investment Trusts (REITs): A viable financing alternative*. https://home.kpmg/in/en/home.html

Le, T. H., & Phan, D. (2023). Blockchain and infrastructure finance: Opportunities, challenges, and policy implications. *Journal of Financial Innovation and Sustainability, 5*(1), 33–51.

Loan Market Association (LMA). (2023). *Sustainability-linked loan principles*. https://www.lma.eu.com

Mazzucato, M., & Semieniuk, G. (2018). Financing renewable energy: Who is financing what and why it matters. *Technological Forecasting and Social Change, 127*, 8–22. https://doi.org/10.1016/j.techfore.2017.05.021

MoRTH (2021). *Annual report 2020–21.* Ministry of Road Transport and Highways, Government of India.

National Infrastructure Investment Fund (NIIF). (2023). *Annual review.* https://www.niifindia.in

National Infrastructure Investment Fund (NIIF). (2023). *Overview of infrastructure investment models in India.* https://www.niifindia.in

Nelson, D., & Shrimali, G. (2014). Finance mechanisms for lowering the cost of renewable energy in rapid deployment markets. *Climate Policy Initiative.* https://climatepolicyinitiative.org

NIIF. (2023). *Annual review: National infrastructure investment fund.* https://www.niifindia.in

OECD & WEF. (2021). *Blended finance for sustainable development: Scaling up private investment.* https://www.oecd.org

OECD. (2021). *Mobilising Institutional Investment in Sustainable Infrastructure.*

OECD. (2023). *Tokenisation of real assets: Opportunities and policy challenges.*

OECD. (2020a). *Developing sustainable finance definitions and taxonomies. Organisation for economic co-operation and development.* https://www.oecd.org/finance/developing-sustainable-finance-definitions-and-taxonomies.pdf

OECD. (2020b). *Developing sustainable infrastructure for a low-carbon future.* https://www.oecd.org/environment

OECD. (2022a). *Infrastructure for a changing world: The role of development finance institutions.* OECD Publishing.

OECD. (2022b). *Sustainability-linked instruments for infrastructure finance.* https://www.oecd.org

Park, S. K. (2018). Investors as regulators: Green bonds and the governance challenges of the sustainable finance revolution. *Stanford Journal of International Law, 54*(1).

PowerGrid (2021). *PowerGrid infrastructure investment trust: Investor presentation.* PowerGrid Corporation of India.

PwC India. (2020). *Infrastructure investment trusts: Unlocking capital for infrastructure development.* https://www.pwc.in

Rodrik, D. (2007). *One economics.* Globalization, Institutions, and Economic Growth. Princeton University Press.

Romero, M. J. (2021). *Public-private partnerships: Defusing the ticking time bomb.* Eurodad. https://www.eurodad.org/public_private_partnerships_defusing_the_ticking_time_bomb

Rozenberg, J., & Fay, M. (2019). *Beyond the gap: How countries can afford the infrastructure they need while protecting the planet.* World Bank. https://doi.org/10.1596/978-1-4648-1363-2

SEBI (2014). *Infrastructure Investment Trusts (InvITs) Regulations*. Securities and Exchange Board of India.
SEBI (2023). *Regulatory framework for infrastructure investment trusts*. Securities and Exchange Board of India.
Securities and Exchange Board of India (SEBI). (2014). SEBI (Infrastructure Investment Trusts) Regulations, 2014. https://www.sebi.gov.in
Sen, A. (1999). *Development as freedom*. Oxford University Press.
Sen, A. (2009). *The idea of justice*. Harvard University Press.
Sharma, P., & Ramani, V. (2023). The ESG mirage? Institutional investors and the politics of climate finance. *Environmental Politics, 32*(1), 155–173. https://doi.org/10.1080/09644016.2022.2130165
Singh, A. (2021). *Infrastructure monetization in India: A policy perspective*. National Institute of Public Finance and Policy.
Stiglitz, J. E., Lin, J. Y., & Monga, C. (2013). *The quality of growth agenda: The case for a new development paradigm*. World Bank.
Tang, D. Y., & Zhang, Y. (2020). Do shareholders benefit from green bonds? *Journal of Corporate Finance, 61*.
TCFD. (2017). *Final report: Recommendations of the task force on climate-related financial disclosures*. https://www.fsb-tcfd.org/publications
Tooze, A. (2021). *Shutdown: How COVID shook the world's economy*. Viking.
UNCTAD. (2023). *World investment report 2023: Investing in sustainable energy for all*. United Nations Conference on Trade and Development. https://unctad.org/wir
UNDP (2022). *Fintech for infrastructure: Innovations from the global south*.
UNDP Ghana (2021). *Blockchain and transparency in public infrastructure*.
UNDP. (2020). *Human development report 2020: The next frontier—Human development and the Anthropocene*. United Nations Development Programme.
UNDP. (2021). *Blockchain for development: Case studies and tools*. United Nations Development Programme. https://www.undp.org
UNEP FI (2023). *Global dtate of Sustainable Finance*.
UNEP FI. (2021). *Net-Zero asset owner alliance progress report*. United Nations Environment Programme Finance Initiative.
United Nations Development Programme (UNDP). (2021). *Blockchain for development: Case studies and tools*. https://www.undp.org
Watkins, K. (2020). *Financing the future: Rethinking infrastructure in a climate-constrained world*. ODI Working Paper. https://odi.org/en/publications
Whitfield, D. (2012). *In place of austerity: Reconstructing the economy*. State and Public Services.
World Bank. (2021). *Incentive-based finance for sustainable infrastructure*. World Bank Group.
World Bank. (2022a). *Blockchain for development: Case studies and lessons*.

World Bank. (2022b). *Capital market innovations for infrastructure financing in emerging economies.*

World Bank. (2022c). *Innovative Financing Mechanisms for Sustainable Infrastructure.* World Bank Group.

World Bank. (2022d). *Green bond impact report.* https://www.worldbank.org

World Economic Forum. (2022). *Blockchain for sustainable infrastructure finance.* https://www.weforum.org

Zerbib, O. D. (2019). The effect of pro-environmental preferences on bond prices: Evidence from green bonds. *Journal of Banking & Finance, 98.*

Zetzsche, D. A., Buckley, R. P., Arner, D. W., & Barberis, J. N. (2020). Decentralized finance. *Journal of Financial, 6*(2), 172–203. https://doi.org/10.1093/jfr/fjaa010

Zhang, B., Wang, Z., & Liu, Y. (2022). Digital finance and sustainable infrastructure: A review of innovations and barriers. *Journal of Environmental Management, 315*, Article 115181. https://doi.org/10.1016/j.jenvman.2022.115181

Zwitter, A., Boisse-Despiaux, M., & Sovacool, B. K. (2020). Blockchain for climate action. *Climate Policy, 20*(9), 1163–1176.

CHAPTER 8

Development of BRICS Energy Cooperation in the Context of Global Challenges and Threats

Ksenia Voronina

1 Introduction

BRICS in many ways exemplifies a strong partnership based on openness and responsiveness. It is important to see the prospects that this partnership offers to the participating countries. The imbalance of the global economy, the rupture and reorientation of ties have led to certain downturns within BRICS, including the weakening of connection between its members, that have started to follow mostly their national interests. This situation points to the need for a certain reforming of BRICS and the development of new approaches. Taking into account Russia's priorities, all these challenges can be grouped into three areas identified as the priorities of the Russian representation. First, there is global stability. The pandemic has led to an escalation of conflicts between countries, as well as to the fact that countries began to close their economies. The

K. Voronina (✉)
Moscow State Institute of International Relations (MGIMO), Moscow, Russia
e-mail: k.voronina@inno.mgimo.ru

N. Tshikovhi et al. (eds.), *The BRICS Imperative*,
https://doi.org/10.1007/978-3-032-05932-1_8

"hybrid war" between the US and China should be noted, the logic of which forces the BRICS countries to define their position, as the risk of unleashing a global conflict is increasing day by day. The sanctions policy of the U.S. and EU against Russia's oil and gas industry (technological and financial sanctions, a price ceiling, a ban on insurance of ships carrying Russian oil) has led Russia to change the direction of its oil and oil products export policy. The largest volume of Russian crude oil supplies currently goes to such BRICS countries as India and China. In 2023 China accounted for about 45–50% of Russian oil exports, and India's share has grown from almost zero to 40% in two years (General Administration of Customs of China, 2023; Ministry of Commerce and Industry of India, 2023).With the collapse of the system for maintaining global strategic stability and the development of regional crises, the organization faces serious challenges to become an "archipelago of stability" in an increasingly unpredictable world.

The BRICS+ institutional mechanism, the formation of which Russia should contribute to, can provide certain opportunities for interaction and increasing both strategic stability and common security which is the second priority for Russia. Russia's third priority is innovative growth. Life dictates new technological priorities: the BRICS countries need flexibility and coordination, especially in connection with the reluctance of the West to exchange technologies. In this regard, it is important not only to cooperate within BRICS itself, but also to develop the BRICS+ format, i.e., multiformat cooperation with other countries. Due to the fact that the energy transition determines the strategic trend of global energy development, the transformation of the energy sector of the BRICS countries is inevitable. As for cooperation between the BRICS countries in the framework of energy transition and in particular the development of green technologies as an innovative component, there are great prospects for the exchange of experience and knowledge, the implementation of joint green projects, and an increase in the level of investment. China currently occupies a leading position in terms of installed renewable energy capacity. Brazil ranks third and India takes the fourth position (Statista, 2023).

Today, the BRICS organization has already established itself as a serious actor with its own institutions and instruments of influence. From the beginning of its existence, when many analysts pointed to the great economic, social, political, and ideological difference between the four, and then—with the accession of South Africa—five countries, they did not take into account the fact that basic approaches to the standards of international law, the source of the legitimacy of governance, regulation of the rules of international interaction for all of them, on the contrary, are close or identical. The approaches of the BRICS countries are characterized by a line of rejection of interference in national sovereignty. Accordingly, the BRICS countries oppose interventionist policies and economic and financial liberalization. Another common position of the BRICS countries is their support for a polycentric world as opposed to a unipolar one. Despite their different status within the United Nations, all BRICS countries, along with multipolarity, the principles of sovereignty and the rule of law, share a commitment to the central role of this organization in modern world governance. There are several areas where important tasks can be solved, thanks to the unification. This includes international trade, economic and investment cooperation, the establishment of a more democratic and fairer international financial and economic order, as well as the use of opportunities to strengthen the export and industrial potential as part of the advantages of interaction in bilateral and five-sided formats.

To date, the BRICS has managed to agree on the so-called energy research platform, the key objectives of which are to carry out information, analysis and research activities in the interests of the five countries on energy development issues, to develop a coordinated energy policy and strategy. The demand for energy is growing with the growing economic development of the BRICS countries, so the development of energy cooperation within the organization is necessary for the further prosperity of the economies of the member countries. If traditional energy sources are now the main sources of electricity generation, then in the medium term their share in energy production will decrease due to the increasing share of renewable energy sources. Of course, a complete abandonment of, for example, oil is impossible due to society's need for petroleum products, so the petrochemical industry will remain in any case. However, it is already clear that the BRICS countries are committed to implementing the energy transition and increasing the share of renewable energy sources in their energy balance. This largely concerns China and India, since they are the main importers of hydrocarbons. Russia's dependence on countries that continue to buy Russian oil and gas is a risk for the Russian economy. In the event of a decrease in purchases of oil and gas from Russia due to the development of renewable energy sources within the BRICS countries, oil and gas revenues from oil and gas exports, which are the main part of the

Russian budget, will decrease significantly. Thus, Russia needs to integrate the project activities of the BRICS countries, which are distinguished by a high degree of implementation of the energy transition policy, and adopt best practices in order to prevent technological lag. Since its first summit in June 2009, BRICS has emerged as an institution of global economic governance and a full-fledged mechanism for strategic and ongoing interaction on key issues of global politics and economy. The agenda of the five includes more than 20 areas of interaction. Despite the fact that critics often call BRICS a political organization, almost 160 commitments (25%) are made in the sphere of economy, trade and finance.

1.1 Methodology

During the writing of the article, a large volume of Russian and foreign literature was studied. The works of the following Russian scientists were analyzed: R.V. Antoshkin (2021), A.M. Mastepanov (2016, 2021), Z.S. Podoba (2018), O.V. Zhdaneev (2021), the subject of which was the replacement of traditional energy sources with renewable ones in the BRICS countries, the role of green investments and features of sustainable development in the context of energy cooperation between the BRICS countries. The works of foreign scientists such as V.G. Azevedo (2018), A. Ferreira (2018), J. Kirton (2018), Y. Wang (2016, 2014) were studied, which dealt with the theoretical and practical foundations of the energy pricing and payment issues in BRICS countries, integration of the BRICS countries in the context of the energy transition to renewable energy sources. Statistical data show the current situation in the Russian energy sector in the context of sanctions from Western countries, energy consumption in the BRICS countries, world greenhouse gas emissions and renewable energy generation in the BRICS countries. Using the methods of generalization and synthesis, the appropriate conclusions were presented.

2 Energy Cooperation Among BRICS Countries

In the context of the EU and US sanctions policy against Russia, Russia continues to strengthen connections with the BRICS countries. Due to the introduction of a price ceiling and embargo on Russian oil and petroleum products, a ban on insurance and reinsurance of ships carrying Russian oil, disconnection from the SWIFT system, restrictions on the supply of equipment, Russia changed the direction of its export policy and logistics routes for the supply of oil and petroleum products, and developed alternative payment systems and payment schemes in the currencies of friendly countries, successfully applied the mechanism of parallel imports. The Russian VTB Bank has developed an alternative payment system for making payments with China. Russia provides discounts on the prices at which it sells oil to China and India. Currently, the main importers of Russian oil among the BRICS countries are China and India. Increasing the level of supplies to these countries is difficult due to the fact that it is not currently possible to lay an additional pipeline to these countries—since the Chinese section of the Eastern Siberia—Pacific Ocean oil pipeline is loaded at maximum capacity, and, as a result, the construction of an additional pipeline will be required. The gas industry seems more promising in this sense, within the framework of which the project for the construction of the "Power of Siberia 2" gas pipeline, connecting the territories of the Russian Federation and China, is being implemented.

Laying pipelines to India is difficult due to factors such as geographical remoteness and geopolitical difficulties. The latter in this case are connected not so much with the previously described geopolitical challenges for the Russian Federation, but with the fact that the construction of the oil pipeline will pass through the territories of Afghanistan and Pakistan, which are unfriendly for India. Brazil has significantly increased the imports of Russian petroleum products after the sanctions were imposed against Russia. As for the Russian gas industry, due to the fact that the gas market operates on the basis of long-term contracts, sanctions have not affected this market to the same extent as they have affected the Russian oil market. Given the technological limitations, Russia cannot fully import foreign equipment that is necessary for the oil and oil refining industries, especially for the extraction of hard-to-recover reserves and deep oil refining. In the current situation, Russia should not rely solely on the mechanism of parallel imports. It is necessary to use the sanctions period to develop domestic technologies with the help of the BRICS countries. China can provide significant support in this regard, for example, in providing some components, despite the fact that China is one of the main oil importers.

According to statistics for 2022, Brazil provides 63% of its electricity generation from hydroelectric power plants, Russia—46% from gas, India—72% from coal, China—63% from coal, South Africa—86%

from coal (World Energy Outlook, 2022). It is worth noting that the price environment for traditional energy sources remains unstable, such resources are exhaustible, the demand for energy is increasing and the issue of renewable energy sources is becoming increasingly relevant. At the moment, the energy transition for many countries, especially for countries exporting hydrocarbons, is perceived as a challenge. Countries are trying to calculate the risks and determine the effectiveness of such a transition. Of course, renewable energy sources have disadvantages—high costs, dependence on climatic conditions, the problem of disposal of renewable energy generation facilities, changes in biodiversity and ecosystems. Now, for example, Russian oil companies use renewable energy generation facilities exclusively for their needs; such electricity does not enter the unified energy system (Rosneft, 2022). However, the energy transition should not be perceived only as a transition to renewable energy sources; the energy transition also includes a technological component—the development of environmentally friendly technologies, oil and gas leak sensors, and others. What is more important for oil and gas exporting countries is to prevent competition between traditional energy sources and renewable ones. Therefore, considering the full picture of renewable energy in the BRICS countries is relevant.

2.1 Development of Renewable Energy Sources in BRICS Countries

Certainly, there are big differences between the BRICS countries in terms of size, socio-economic conditions and the use of energy resources. In the near future, China will remain a major importer of oil, while Brazil will become a major exporter of crude oil. Nevertheless, both China and Brazil are making optimistic plans on how they will realize the need to follow the path to a sustainable energy transition. This offers great opportunities for improving energy cooperation between China and Brazil. Energy cooperation between these countries should embrace their innovative and technological capabilities to facilitate the transition from fossil fuels to renewable energy in the mid-term. China's energy policy is developing through a combination of central control exercised by the government administration and investment plans for private and public energy companies operating in the country. Fossil fuels account for about two-thirds of the primary energy used in China. Although China plans to increase the share of renewables in its energy balance, especially the share of wind and solar power, fossil fuel consumption has increased rapidly over the past few years and the trend is not changing. China has a relatively small hydrocarbon reserve. As a result, unable to find its own deposits to meet national demand, China is seeking to strengthen cooperation within its geographic region, considering Russia as its key partner in energy imports (Energy Review of the BRICS Countries, 2020).

India's conceptual approach to ensuring economic growth based on its own resources involves solving the tasks of reducing the volume of energy imports, electrifying the population and sustainable development. All three tasks are solved through an increase in the share of renewable energy sources in the fuel and energy balance with the localization of the best available technologies. The established National Solar Mission and The International Solar Alliance are central drivers of electrification in India. South Africa's strategic approach to energy is based on the established tradition of relying on its own technologies and resources. Modern challenges, including those related to the tightening of the environmental agenda, determine the relevance of the search for options to implement measures to create new production segments based on the competitive advantages of an efficient mining industry. Extensive deposits of platinum group metals (leading in the world in terms of platinum and iridium production) and the potential of an electricity base based on renewable energy sources are key prerequisites for creating an even more powerful energy system in South Africa.

The focus on only one type of energy will automatically lead to lower energy security. Diversification of the fuel and energy balance mitigates the effects of single-source supply disruptions by offering alternatives that meet the interests of both consumers and producers, for whom market stability is of paramount importance. Both developed and developing countries have begun the process of diversifying energy sources mainly through the development and utilization of nuclear, wind, solar, hydropower and other new forms of energy. This idea has been widely supported by the International Energy Agency, which emphasizes the critical importance of the transition to modern fuels for a country's future economic development. The criteria for assessing the efficiency of a country/region in such a transition are the Energy Development Index (EDI) and Energy Efficiency Index (EEI).

2.2 *Cooperation in the Field of Energy Security of the BRICS Countries*

Since 2009, energy security has been on the agenda of BRICS summits and ministerial meetings. The BRICS countries' interaction in the field of energy security is characterized by the similarity of their approaches to international cooperation, the proximity of their domestic socio-economic development objectives, and the need to find answers to common global challenges faced by the countries (Statement of the 6th BRICS environment ministers meeting, 2020).

The BRICS countries are parties to all major international agreements on energy and climate, including the Paris Agreement and the UN Framework Convention on Climate Change. India, Russia, and China are among the countries with the highest specific greenhouse gas emissions. According to the world statistics, the top five countries emitting GHGs are China, the United States, India, Russia, and Japan. They account for over 60% of global emissions (IEA CO2 Emissions in 2022). The corresponding chart is presented below (Fig. 1).

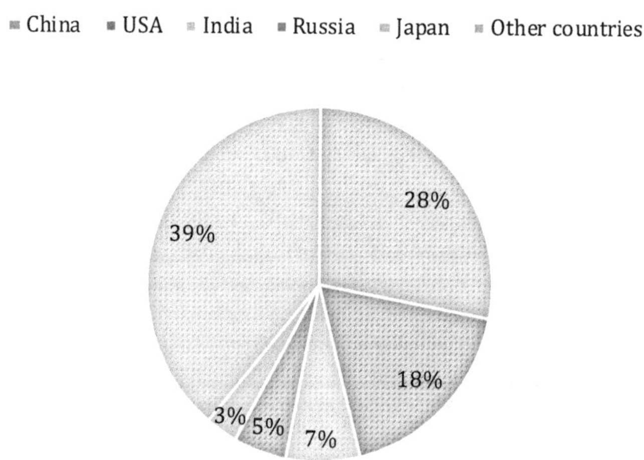

Fig. 1 World greenhouse gas emissions (%), 2022 (*Source* Compiled by the author based on [IEA CO2 Emissions in 2022])

In 2015, at the first meeting of BRICS energy ministers held in Russia, a working group on energy saving and energy efficiency was established. The main issues discussed by the Heads of State and ministers were energy efficiency improvement, diversification of energy sources, application of new environmentally friendly technologies, and development of scientific and technical cooperation. Progressive work is underway in all these areas.

2.3 *Current Energy Cooperation between the BRICS Countries*

Despite the common priorities outlined in the BRICS official documents, there are some differences between the energy systems of the BRICS countries and their national interests. First of all, these are structural differences in the countries' energy balances, in the volume of energy consumption (Fig. 2) and in energy transportation systems.

The group of countries includes both large net exporters (Russia and Brazil) and net importers (China and India). China and India are the main drivers of global energy demand due to their growing economies and level of population. China has been the world's largest energy consumer for several years, and its energy demand in 2040 will be twice that of the US.

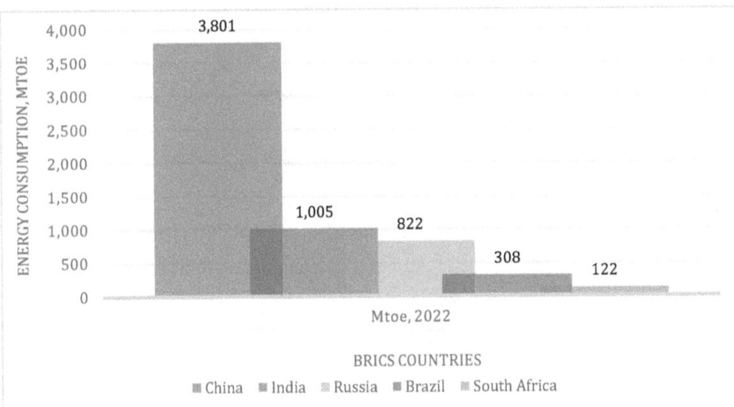

Fig. 2 Energy consumption in the BRICS countries (Mtoe), 2022 (*Source* Compiled by the author based on [IAE World Energy Outlook, 2022])

Despite the absolute growth in the consumption of all types of energy resources, including thermal coal, the energy balance of the BRICS countries is becoming more and more green. All five countries are increasing their energy generation from renewable sources. China and Brazil are leading in this area, both in terms of investment in renewable energy and installed capacity of renewable energy power plants. The full picture of renewable energy generation in the BRICS countries is presented in the Fig. 3.

The active establishment of the new industry and the achievement of price parity with fossil fuels allow the authorities to gradually reduce subsidies from the state budget. It is renewable energy that has been declared as a priority topic of BRICS scientific and technological cooperation, for which China is responsible. Cooperation in this area is also developing in a bilateral format, for example, through the work of the Russian-Chinese Working Group on Cooperation in the Use of Renewable Energy Sources.

Despite the short history of Russian renewable energy, Russia already has a lot to offer its BRICS partners. A landmark project was the construction of the first phase of solar power plants in South Africa, carried out by a subsidiary of "Renova" Group for the South African holding Dawn Group. Research in the energy sector is among the priorities of BRICS scientific and technical cooperation. This is confirmed not only by the

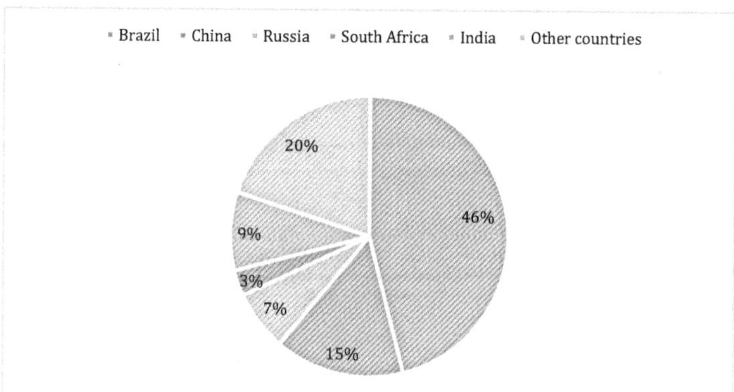

Fig. 3 Renewable energy generation in the BRICS countries (%), 2022 (*Source* Compiled by the author based on [IAE World Energy Outlook, 2022])

topics of multilateral competitions of scientific and technical projects organized by the BRICS Framework Program, but also by the number of publications in this area (BRICS STI Framework Programme, 2019). BRICS energy cooperation is based on international agreements and long-term contracts. The most intensive cooperation is in the nuclear energy sector, including the provision of services for the construction and maintenance of nuclear power plants, supply of fuel for nuclear power plants, waste treatment and training of specialists. For example, cooperation between Russia and China in the nuclear energy sector is developing in a wide range of areas, including the construction of demonstration and operating power units, the creation of infrastructure for nuclear fuel cycle facilities, the supply of fuel. Russia is among India's key partners in the peaceful use of nuclear energy, with the first agreements signed in the late 1980s. In 2018. Russia and India signed the "Action Plan for Prioritization and Implementation of Cooperation Areas in the Nuclear Field", according to which it is planned to build six nuclear power plant units in India. In 2017 Russian State Atomic Energy Corporation "Rosatom" signed a memorandum of understanding with Brazilian companies "Eletrobras" and "Eletronuclear" to develop cooperation in the peaceful use of atomic energy. In 2014, Russia and South Africa signed an intergovernmental agreement on strategic partnership in nuclear energy and industry. Progress in this area of cooperation is noticeable between Russia, China, and India, while South Africa and Brazil have so far refused to further jointly develop nuclear energy due to financial difficulties.

The BRICS countries are interested in creating new institutions and strengthening cooperation in the energy sector to ensure stability of demand and supply and reduce volatility of energy prices. Such cooperation makes it possible to compensate for the deterioration in relations of each of the BRICS countries with third countries. For example, the imposition of sanctions against Russian oil and gas companies in 2014 forced them to partially reorient themselves to new markets (primarily China and India), increasing oil and coal exports. New projects in this area include Rosneft's participation in the construction of a refining and petrochemical complex in China and preparations for the development of oil and gas fields in Brazil. At the same time, in the context of sanctions against Russia, China got an opportunity to take part in the implementation of projects in Russia, to offer its competencies and equipment in geological exploration, production, processing and transportation of hydrocarbons.

3 Discussion

The priorities for the near future are: first of all, the development of national energy systems of the BRICS countries, the development of energy cooperation between the countries, the introduction of new technologies and the improvement of conditions for investment in the energy sector. A special mention should be made of the expansion of the BRICS Energy Research Platform. This Russian initiative was developed in accordance with the seventh UN Sustainable Development Goal and makes it possible to implement the priorities enshrined in the BRICS documents and decisions. Conducting research in priority areas for countries involves constant monitoring of global trends and analysis of their impact on the BRICS countries.

It is advisable to involve private sector organizations in the implementation of the BRICS priority of universal access to energy, its affordability, pollution reduction and environmental protection, which requires international cooperation and is in line with the seventh UN Sustainable Development Goal. Such projects could be financed by the BRICS New Development Bank, which has already financed 10 clean energy projects. In addition, cooperation with international organizations with which the BRICS countries have already gained positive experience of cooperation, including the World Energy Council (WEC) and national WEC committees in the BRICS countries, will be reasonable.

According to the BRICS documents, cooperation activities to enhance energy security can be grouped into four areas:

1. Identification of approaches to diversification of energy resources, including renewable and low-carbon energy sources;
2. Improvement of mechanisms for attracting investments in the energy sector and infrastructure;
3. Elaboration of strategies to ensure universal access to energy and its affordability;
4. Development of technologies to reduce the negative impact of the energy industry on the environment and climate.

These areas can be primarily explored in the course of prospective studies using foresight tools and involving interested scientists who are members of the Energy Research Platform. In the future, the identified priorities can be studied in more detail in the format of applied research projects involving discussions at public and expert events, development of action plans, sectoral projects, development assistance programs and new tools for cooperation between the five countries.

The Paris Climate Agreement outlined a global consensus on the need for a transition to a low-carbon future. Environmental issues occupy a key place in the Sustainable Development Goals set by countries on the horizon until 2030. Environmental quality is becoming one of the most important criteria for successful economic development. The concepts of "green economic growth", "green finance", "green jobs" are now found not only in the releases of environmental projects, but also in the reports of international organizations, in the official strategies of leading states, in the programs of political parties, in the reports of the world's largest companies and banks. Many of them declare the transition to a "green economy" as a new development paradigm that envisions increased prosperity and social justice while reducing environmental risks and conserving natural capital. While some consider the green economy to be impossible to achieve on a global scale, the "green transformation" is gaining momentum: the energy sector is rapidly shifting to less carbon-intensive energy sources; radical changes are taking place in the automotive and construction industries; land, water, and forest management practices are changing. Most importantly, companies' approach to the environment is changing: the management of its elements is becoming an essential part of the business. These changes are caused, on the one hand, by the aggravation of environmental problems associated with anthropogenic pressure on the environment, and, on the other hand, by a change in the attitude of states and companies to them, and by the close intertwining of environmental problems with economic ones.

In the environmental sphere, the BRICS countries share a number of common features. Firstly, their ecosystems are unique. For example, Russia and Brazil together account for more than a third of the world's forests. In Russia there is a unique ecosystem of Lake Baikal, in Brazil—the ecosystem of the Amazon River. China and India are the largest countries with the Himalayan Mountains and their glaciers play a key role in solving the problem of water shortage, providing water for about 2 billion people. Russia accounts for most of the unique Arctic zone, which is particularly vulnerable to climate change.

Secondly, these countries are extremely ecologically unfavorable. Their positions in the ranking of countries on the Environmental Performance Index, compiled on a regular basis by Yale University and summarizing the whole range of environmental problems faced by the countries, are indicative. All BRICS countries are characterized to varying degrees by high levels of air, water, and soil pollution (especially in China and India), waste disposal problems (in all countries), deforestation and forest fires (especially in Russia and Brazil), biodiversity loss (especially in Brazil and South Africa), water scarcity (especially in India, China, and South Africa), etc. Environmental problems have a huge negative impact on public health, causing additional morbidity and mortality, as well as on the state of the economy.

Thirdly, in all BRICS countries, environmental degradation is closely embedded in current economic growth patterns. The economic development of these countries is still largely extensive, due to the involvement of more and more natural resources in the economic turnover.

The environmental problems of the BRICS countries are also closely linked to the structure of their energy balance. In China, India, and South Africa, the dirtiest fossil fuel—coal—completely dominates the energy balance. While in China and South Africa the energy balance will become cleaner in the future (coal will gradually be replaced by natural gas, renewables, and nuclear power), the situation in India is rather pessimistic. Russia's primary energy consumption structure is much cleaner, but the country specializes in exporting fossil fuels, and significant pollution is generated at the stage of its production and transportation.

4 Conclusion

Russia does not have a developed domestic infrastructure in the field of renewable energy development; most project activities are carried out abroad, renewable energy generation facilities are exported. This situation casts doubt on the further development of renewable energy sources, since, according to the experience of other countries that have a fairly high share of renewable energy sources in the energy balance, the primary task was precisely the creation of a domestic market and then entering foreign ones. However, in Russia the domestic market remains dominated by fossil fuel producers. It is important for Russia to avoid such a dilemma when it has to choose between the green agenda and traditional energy, so it is necessary to join forces in this direction and adhere to coordinated actions on both aspects (Makarov, 2020).

The sanctions policy against Russia strengthened the nature of ties within the BRICS countries—there was an increase in oil supplies to

China and India, petroleum products to Brazil, alternative payment systems were developed with friendly countries, payments began to be made in the currencies of friendly countries (the role of the Chinese yuan and Indian rupee increased). The stable continuation of Russian oil supplies to friendly countries may be affected by new packages of sanctions from the US and EU, which are aimed at ensuring that friendly countries also stop buying oil from Russia, like Western countries. This situation can already be observed with Russian oil tankers that are idling near the coast of India. Russia believes that the BRICS countries will continue mutually beneficial cooperation in achieving a common goal—maintaining a multipolar world. Renewable energy sources are a strategic trend in the global energy sector, so strengthening cooperation in this direction and reducing countries' dependence on traditional energy sources is relevant.

In their joint statements at the end of each summit, the BRICS leaders express their commitment to key international agreements on the environment and sustainable development, including the Rio Convention on Sustainable Development, the Sustainable Development Goals, the Paris Climate Agreement, the Convention on Biological Diversity, etc. Leaders also express their willingness to cooperate in addressing relevant environmental issues with other countries within the framework of existing international mechanisms. The BRICS countries seek to continue to promote green development and low-carbon economy in the context of sustainable development and poverty eradication, enhance BRICS cooperation on climate change and expand green finance. At the same time, while recognizing the readiness to combat climate change, the BRICS countries point to the need to adhere to common but differentiated responsibilities and call on developed countries to provide financial, technological and capacity building support to developing countries to enhance their ability to mitigate and adapt to climate change. Since 2015, BRICS environment ministers have met annually to agree not only on general principles of cooperation in this area, but also on specific initiatives. In particular, agreements were reached on technical expertise in the field of air and water pollution control. As a result of the meeting of environment ministers in São Paulo in 2019, it was decided to create a list of the best available technologies for industry.

Environmental cooperation also covers the work of the BRICS New Development Bank. "Green economy" is one of the priority areas of project financing by this organization. Compliance with the principles of sustainable development is also an important criterion in the selection of infrastructure and energy projects that receive financing from the bank. Environmental issues will continue to occupy an important place in BRICS cooperation. The implementation of already adopted initiatives will continue, as well as the financing of green projects by the BRICS New Development Bank. Despite the ambitious plans and the large amount of work already done, it seems that the potential for cooperation in the field of environment and climate change in the BRICS countries could be much higher. Narrow initiatives designed to ensure the exchange of experience on specific environmental issues, as well as the financing of individual green projects, are important but insufficient areas of cooperation (Morozkina, 2015). The BRICS countries should take a broader view of environmental issues—in the context of their economic growth potential and their role in the world. The BRICS countries are the first to move from low to middle and high per capita incomes in a world of global climate and environmental crisis. Due to resource and environmental constraints, they will not be able to achieve this transition by duplicating the growth model of Western countries based on Western consumer values. Although the green development debate is currently taking place predominantly in the rich world, it is for the BRICS countries that the green growth model is a necessity rather than an act of civic consciousness. This necessity is emphasized by the exhaustion of the potential of their current growth models, manifested in the declining growth rates of the BRICS economies and, in some countries, in stagnation.

Discussion of new ecological growth models that take into account resource and environmental constraints, as well as the general trend toward green transformation of the global economy, could become an important area of cooperation within the BRICS framework. Moreover, individual countries of the organization have their own ideas about this. In particular, in 2012 China proposed the idea of "ecological civilization" and developed the "Strategy for Energy Revolution (2016–2030)". South Africa also has experience in developing detailed national plans for "green development". It is important that possible success in developing new models of growth will immediately make the BRICS countries a role model for the states that follow them on the path of development: from developing giants like Indonesia, Pakistan or Nigeria to small countries in Africa and Asia.

The discussion of new growth models will inevitably raise the question for the BRICS countries about their role in solving global environmental problems, primarily global climate change. Today, the international regime in this area is formed mainly by developed countries (which account for only one third of global emissions) and based on their vision of the world. For example, this regime takes into account those emissions that are produced within a country's territory and ignores emissions from international trade flows. This allows developed countries to succeed in reducing emissions without reducing their consumption, only by replacing the production of their own carbon-intensive goods with their imports. These imports come mainly from the four BRICS countries: China, Russia, India, and South Africa.

In contrast, the alternative accounting for emissions—by consumption—only includes emissions from the production of goods consumed in the country (including emissions from imports). The shift to consumption-based emissions accounting fundamentally changes the view of the causes of emissions growth, shifting the focus from producers of fossil fuels and carbon-intensive goods to their consumers, whose demand is the root cause of emissions (United Nations Framework Convention on Climate Change, 2022). This approach does not allow the BRICS countries and their specialization to be seen as the main challenge to low-carbon growth, but asks how to curb the uncontrolled consumption growth of the rich segments of the world's population. The same question is relevant to the BRICS countries' search for new models of inclusive green growth, free from the goals of maximizing consumption. Against the backdrop of increasing attempts to politically and internationally isolate Russia by the West, it is recommended to intensify BRICS contacts and make them more regular, formal and systematic, as Russia needs platforms to defend its global interests. Russia should continue to adhere to the principles of avoiding double standards and respect for international law shared by all BRICS member states. The optimal option for BRICS development is to further expand the network format of integrated political coordination.

Based on the New Development Bank of BRICS, it is necessary to develop a new format of the international payment system for mutual settlements and to overcome the negative impact of unilateral anti-Russian measures by a number of Western countries in order to provide a basis for stable trade and technology exchange. On the basis of the

New BRICS Development Bank, it is necessary to encourage preferential lending, and switch to settlements in the national currencies of the BRICS members, which will help increase exports and imports.

BRICS countries should be considered separately as markets and additional ways to develop foreign trade. This implies the development of logistics chains and the replacement of foreign companies that have left the Russian market with domestic firms or firms from friendly countries. This will require the formation of the necessary legal framework and institutional platforms for interaction, which will be approved at the meeting of the relevant ministers.

References

Antoshkin R. V. (2021). Development of renewable energy in the interaction of BRICS countries. *Innovations. The science. Education*, (28), 1403–1409.

Azevedo V. G., Sartori, S., & Campos L. M. S. (2018). CO2 emissions: A quantitative analysis among the BRICS nations. *Renewable and Sustainable Energy Reviews*, *81*, 1, 107–115.

BRICS official documents and meetings. Environmental Ministers. Statement of the 6th BRICS environment ministers meeting. 30 July, 2020.—[Electronic resource]. http://www.brics.utoronto.ca/docs/200730-environment.pdf (Access date: 13 April 2023).

BRICS STI Framework Programme. (2019). [Electronic resource]. http://brics-sti.org/ (Access date: 19 April 2023).

CO2 Emissions in 2022. IEA. (2022). [Electronic source]. https://www.iea.org/reports/co2-emissions-in-2022 (Access date: 23 April 2023).

Development prospects and strategy. Rosneft (2022). [Electronic resource]. https://www.rosneft.ru/about/strategy/ (Access date: 27 April 2023).

Energy Review of the BRICS Countries. Energy research cooperation platform. (2020). [Electronic source]. https://brics-russia2020.ru/images/114/89/1148936.pdf (Access date: 20 April 2023).

Ferreira, A., Kunh, S., Fagnani, K., De Souza, T., Tonezer, C., Dos Santos, G., & Coimbra-Araújo, C. (2018). Economic overview of the use and production of photovoltaic solar energy in Brazil. *Renewable and Sustainable Energy Reviews*, *81*, 181–191.

inistration of Customs of China. (2023). [Electronic resource]. http://english.customs.gov.cn/statics/report/monthly.html (Access date: 20 April 2023).

Kirton, J. (2018). *BRICS and global governance* (J. Kirton, & M. Larionova, Eds., p. 288). Routledge.

Leading countries in installed renewable energy capacity worldwide in 2023. Statista. (2023). [Electronic source]. https://www.statista.com/statistics/267233/renewable-energy-capacity-worldwide-by-country/ (Access date: 21 April 2023).

Makarov, I. (2020).Green transformation of the world economy: Risks and opportunities for Russia. In G. Diesen & A. Lukin (Eds.), *Russia in a changing world*. Palgrave Macmillan.

Mastepanov, A. (2016). Cooperation between the BRICS countries in the energy sector as a factor in forecasting global energy consumption. *Drilling and Oil* (1), 3–9.

Mastepanov, A., & Sumin, A. (2021). Energy policy of Brazil. *Energy policy*, (3), 58–79.

Ministry of Commerce and Industry of India. (2023). [Electronic resource]. https://commerce.gov.in/trade-statistics/ (Access date: 17 April 2023).

Morozkina, A. (2015). New Development Bank in the global financial and economic architecture. *International Organisations Research Journal (IORJ)*, *10*(2).

Net-zero carbon dioxide emissions. United Nations Framework Convention on Climate Change. (2022). [Electronic source]. https://unfccc.int/ru (Access date: 25 April 2023).

Podoba, Z. S., & Kryshneva, D. A. (2018). "Green energy" in the BRICS countries. *World Economy And International Relations*, *62*(2), 17–27.

Wang, Y., & Zhi, Q. (2016). The role of green finance in environmental protection: Two aspects of market mechanism and policies. *Energy Procedia*, (104), 311–316.

Wang, Z., He, H., & Fan, M. (2014, November 1). The ecological civilization debate in China: The role of ecological marxism and constructive postmodernism—Beyond the Predicament of Legislation. *Monthly Review*, *66*(6).

Zhdaneev, O. V., & Seregina, A. A. (2021). Vectors of BRICS technological cooperation in the fuel and energy complex (part 2). *International Technical and Economic Journal*, (2), 7–22.

World Energy Outlook. (2022). IAE.—[Electronic source]. https://www.iea.org/reports/world-energy-outlook-2022 (Access date: 25 April 2023).

CHAPTER 9

Prioritizing Sustainable Growth: Trends and Prospects of Addressing Climate Challenges Within BRICS

Filip Selamovski and Svetlana Evgenyevna Sobenina

1 Introduction

Sustainable growth refers to economic expansion that can be maintained in perpetuity without creating other substantial economic problems. Initially used in terms of business operations efficiency, nowadays, the concept of sustainable growth primarily denotes economic advancement that avoids adverse environmental impacts. The result of this sustainable growth and sustainable economy should be a society where all human needs are met while simultaneously preserving the stability of the natural

F. Selamovski
Department of International Business, Plekhanov Russian University of Economics, Moscow, Russia
e-mail: Selamovski.F@rea.ru

S. E. Sobenina (✉)
Special Faculty of Talented Youth "Foresight", Plekhanov Russian University of Economics, Moscow, Russia
e-mail: lana.sobenina@mail.ru

N. Tshikovhi et al. (eds.), *The BRICS Imperative*,
https://doi.org/10.1007/978-3-032-05932-1_9

system. On the background of rising population, accelerating urbanization and climate emergency, achieving sustainable growth has become the number one priority of most of the global economies (Ruggerio, 2021). And BRICS countries are not an exception. Addressing climate challenges within the policy of prioritizing sustainable growth is of great importance for BRICS, given that the need for continuous modernization and structural transformation of their economy, combined with the fact that BRICS countries already account for more than 40% of global greenhouse gas emissions (Tian Xu et al., 2020), could additionally increase emissions and yield negative economic consequences. Therefore, indeed there is a strong course of prioritizing sustainable growth, and addressing climate challenges in particular, among BRICS countries.

Hence, the importance of the question: What are the current trends and prospects of addressing climate challenges within BRICS? This question is addressed by analyzing the bilateral and multilateral mechanisms within BRICS, including the transition from one to another as an emerging trend. Also, divergence among BRICS countries in terms of priorities and action vectors is discussed. Especially, this chapter addresses the issue of importing carbon emissions (carbon leakage) as one of the main challenges BRICS countries face, specifically how they address carbon leakage and try to close the gap between them and developed countries. The chapter is structured in the following way. Firstly, it presents a discussion on the use of bilateral and multilateral mechanisms in the context of addressing climate challenges. Secondly, it analyzes the level of differentiation and disparity among BRICS members, as well as divergence in terms of policy goals and key areas. Thirdly, it presents insight into the practice of "importing" carbon emissions and carbon leakage and the efforts BRICS countries are undertaking in order to address this issue. Lastly, the chapter concludes with a summary of the key findings obtained during the research and a brief discussion on its limitations.

2 Transition from Bilateralism to Multilateralism

2.1 Current Approaches to Addressing Climate Challenges

Since its formation, BRICS countries have acknowledged climate challenges and the necessity to achieve sustainable growth as one of their main priorities. As a group, BRICS countries are strong supporters of the multilateral approach and highlight the need to adhere to the principles of the United Nations Framework Convention on Climate Change and its Paris Agreement (BRICS PNIR, 2023). Although they support multilateralism as an important mechanism to address global issues such as climate challenges, evidence shows that actions within the group are mostly limited to bilateral discussions of the technology exchange possibilities and joint financing of green projects. This is in line with the fact that countries advocate for the global principle of common but differentiated responsibilities, which recognizes the differing capacities. A great example of bilateralism's preferability are the ongoing efforts of China and Brazil to establish a bilateral fund for financing the development of green industry and renewable energy as an instrument to combat climate challenges and achieve sustainable growth. Additionally, both countries created a special subcommittee on climate change within the Sino-Brazilian High-level Commission for Consultation and Cooperation to cooperate on policy and technical issues (MFA of Brazil, 2023). Simultaneously, China and Russia have been substantially increasing mutual cooperation in terms of project proposals and mobilizing resources, including financial ones, in the field of green development (People's Daily, 2023).

Over the years, they have undertaken jointly-funded actions to tackle air and water pollution, preserve endangered wildlife and promote green and sustainable development overall. Revisiting the preference for bilateralism, it is worth noting that the aforementioned interactions between member states could conceivably occur, either partially or entirely, via the New Development Bank (NDB)—a multilateral financial organization founded by BRICS nations with a mission to fund sustainable development projects. However, the decision to adhere to bilateral arrangements and bypass an existing functional platform for cross-border coordination, including addressing climate-related issues, underscores a lingering reluctance among member states to expand specific initiatives into multilateral formats. This reticence likely stems from multiple factors, including skepticism about the effectiveness of centralized governance structures. Nations may prioritize customizing initiatives according to their unique circumstances, preferring ad hoc collaborations over institutionalized mechanisms. Additionally, the absence of universally agreed-upon standards or enforcement mechanisms within multilateral settings adds complexity, deterring participation in broader frameworks.

Nonetheless, this hesitation leaves numerous untapped opportunities for collective problem-solving and synergistic gains, as demonstrated by the potential offered by the NDB. Addressing these reservations requires deliberate efforts to build long-term trust in its efficacy, clarify shared objectives, and construct resilient multilateral institutions capable of bridging divides and fostering unity. To a certain extent, the prevalence of bilateralism within the BRICS framework can be regarded as reasonable and logical. This argument rests on several foundational premises. Firstly, the member nations of BRICS confront diverse and disparate climate-related challenges, which naturally dictate their individualized approaches to environmental policymaking. Secondly, unlike supranational bodies advocating deep economic or political unification, BRICS rather emphasizes a multidimensional strategic partnership anchored in shared values and mutual respect (Kirpizli, 2022; Ruppel & Ruppel Schlichting, 2013).

Nevertheless, a new trend of transition from bilateralism to multilateralism within BRICS has been slowly emerging. An initiative called BRICS Resource Efficiency and Circular Economy Dialogue has been initiated by India with the aim of promoting the sharing of knowledge and best practices in waste management, resource efficiency, and circular economy as a way to mitigate and avoid further negative environmental impact. Within the scope of this Initiative, the participating countries agreed to engage in discussions pertaining to various sectors including construction, agriculture, solar energy, biofuels, packaging, electronic waste, food, water, and textiles and to shape a common approach accordingly. Moreover, in August 2021, space agencies of BRICS countries signed an agreement on sharing remote sensing satellite data to help cope with climate changes and major disasters and in environmental protection (SC PRC, 2021).

2.2 *Bilateralism v. Multilateralism: Prospective Pathways*

Opinions on which option, bilateralism or multilateralism, is more suitable in the perspective are quite divided. On the one hand, in the context of rising global uncertainty, it is evident that unilateral and even bilateral actions have little to no result. We live in a globalized world, where one country's problems easily become regional if not global problems. Climate challenges are perhaps one of the most important threat that the global economy and society face given they cannot be locally contained, which in line with previous studies showing that multilateralism is a better solution for economic problems that require fully coordinated global efforts (Reich, 2009), underlines the multilateral approach's superiority.

On the other hand, theoretical research argues that multilateralism needs the following three features: indivisibility, generalized organizing principles, and diffuse reciprocity (Ruggie, 1993). Also, it is argued that multilateralism rejects the practice of making distinctions based on specific circumstances, personal influence, or individual requests and requirements (Webber, 1991). Given that BRICS members face different climate challenges due to geographic specifics and different nature of their national economies, there might be some justifications to the statement that BRICS countries are not yet fully compatible with multilateral cooperation in terms of climate policy. In favor of bilateralism is the fact that it reduces costs and accelerates implementation of projects in targeted areas. Bilateralism enables more flexible and tailored solutions, accommodating the unique circumstances faced by each partner. They avoid the pitfalls of rigid hierarchies or standardized regulations that might hinder smaller or less-developed members. Furthermore, bilateral interactions promote trust-building, confidence, and reciprocity, essential elements for constructing long-term partnerships. Therefore, to a certain extent, indeed, viewing bilateralism as a viable interim solution until more sophisticated multilateral mechanisms mature seems prudent. It ensures that core national interests are preserved while allowing room for iterative improvement and eventual evolution toward broader cooperation.

Overall, the potential for intensifying multilateral collaboration spans a wide spectrum, encompassing vital areas such as data exchange, joint ventures in infrastructure development, clean energy breakthroughs, and low-emission production techniques. While ultimate decisions rest solely with individual BRICS nations, informed by their respective strategic objectives, current indicators strongly suggest that multilateral cooperation will increasingly supplant bilateralism as the preferred mode of intra-group interaction. Multilateral platforms hold significant appeal due to their capacity to pool resources, diffuse risks, and catalyze collective action. Through collaborative projects, member states can overcome barriers posed by asymmetric capabilities, divergent priorities, and uneven starting points. Simultaneously, shared responsibilities and accountability mechanisms reinforce transparency, build trust, and promote equitable burden-sharing. Moreover, such frameworks provide a structured environment for reconciling divergent views, generating compromise solutions, and avoiding conflicts over scarce resources. Thus, envisioning a future dominated by multilateralism within BRICS aligns with broader global trends emphasizing interdependence, inclusivity, and solidarity.

3 Divergence in Policy Goals and Actions Among BRICS Members

3.1 Heterogenous Progress Toward Sustainability

As discussed in the previous section, BRICS countries are strong supporters of multilateral and common approaches to key challenges including the formation of a common vision. However, there is an obvious situation of growing levels of disparity in development vectors, as well as the achieved results among them. This lays the foundation to claim that their policy goals and visions after all are not fully aligned and are even furtherly diverging. To substantiate this claim we will first discuss renewable energy as the key area of arising differentiation. Although one primary focal point in tackling climatic issues revolves around harnessing renewable energy sources, BRICS nations exhibit noticeable heterogeneity in terms of advancement within this domain. This disparity substantively shapes their outlooks regarding forthcoming evolutionary trajectories and corresponding normative responses. Evidence of the latter statement is presented in Fig. 1, which illustrates the shares of renewable energy in the total energy supply (TES) and electricity generation in BRICS countries. Analysis of available data underscores a striking dichotomy: Despite China dominating worldwide expenditure directed toward transitionary energy paradigms accounting for approximately 35% of cumulative global investment, equivalent to USD 266 billion in 2021—it paradoxically trails considerably behind Brazil, whose investment fraction constitutes merely 1.6% of planet-wide allocations dedicated to sustainable energy transitions (WEF, 2022). This incongruity highlights divergent trajectories within BRICS, emphasizing unique factors influencing each nation's path toward sustainable development. Factors such as geographical endowments, socio-economic structure, technological prowess, governance mechanisms, and policy inclinations interact dynamically, engendering distinctive trajectories of renewable energy development.

Brazil distinguishes itself prominently, assuming leadership positions not solely within BRICS but internationally in the context of renewable energy utilization. Nearly 46% of Brazil's energy originates from sustainable sources, notably bioenergy and hydropower, positioning it among the least carbon-dependent economies globally (IRENA, 2022). Beyond this, Brazil is actively exploring green hydrogen as a promising low-carbon energy source, complemented by substantial commitments to research and development initiatives focusing on diversifying its energy technology landscape. These efforts underscore Brazil's proactive stance in addressing climate mitigation objectives and securing its place at the forefront of innovative energy solutions.

Among the BRICS nations, Russia and South Africa emerge as apparent outliers concerning the incorporation of renewable energy sources within their respective TES. Specifically, these two countries exhibit relatively minimal shares of renewable energy, accounting for a mere 3% and 7%, respectively, within their overall energy matrices. This discrepancy may be attributed, in part, to their pronounced reliance on

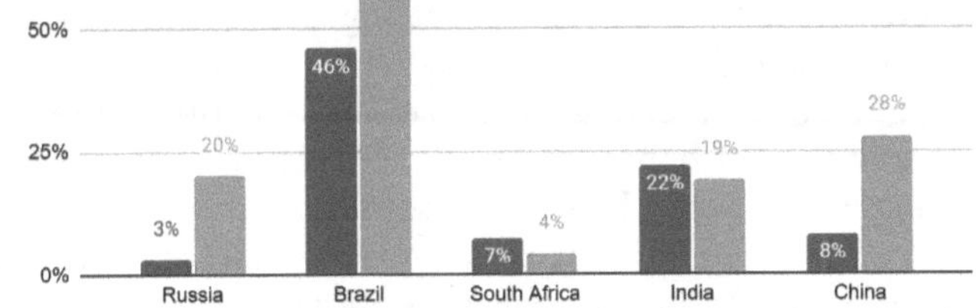

Fig. 1 Share of renewables in the total energy supply and electricity generation in BRICS countries (2004 & 2022) (*Source* International Renewable Energy Agency [2023])

non-renewable forms of energy, namely fossil fuels, which dominate their energy landscapes. For instance, in Russia, oil and natural gas constitute critical contributors to state finances, comprising nearly 42% of federal budget receipts in 2022 (MoF RF, 2023). This dependency inherently dampens motivations for swift transitions toward alternative, cleaner energy sources, creating structural impediments to rapid decarbonization efforts. Similarly, South Africa exhibits a strong propensity toward coal utilization, with this resource serving as the predominant energy input, accounting for approximately 72% of its TES (IRENA, 2022). Coal mining and processing play crucial roles in sustaining economic activity and employment, entrenching the country's reliance on traditional, carbon-intensive energy systems. Consequently, any shift toward greater reliance on renewable energy in these countries entails navigating complex tradeoffs involving energy security, job displacement, and macroeconomic stability.

3.2 Evaluating Divergence: Policy Implications

Having assessed the disparity in progress, we continue to investigate the emerging trend of divergence in current policy goals and actions, arising from this disparity. In order to substantiate this observation, we examine fossil fuels and particularly coal consumption policy and key actions within as suitable example of this underlying tendency. China, India, and Russia exemplify a hybrid approach to energy transition, blending investments in renewable technologies with persistent reliance on fossil fuels. Recent studies indicate that China commissioned six-fold more new coal plants than other nations in 2023, despite leading global renewable energy investments (CREA, 2023). Driven partly by the 2022–2023 energy crisis characterized by scarcity and escalating energy costs, this strategy mostly reflects Beijing's pragmatic concern for energy self-sufficiency, macroeconomic stability and general national security. Nonetheless, it contradicts global decarbonization goals, potentially jeopardizing international climate ambitions. Despite possessing vast untapped renewable energy reserves owing to its immense landmass, Russia remains hesitant to diminish its dependence on non-renewable energy sources. Though incremental efforts have been initiated to exploit this latent potential through varied incentive schemes, projections under the prevailing energy strategy anticipate that renewable energy will comprise only 4.9% of the TES by 2030. This modest target falls far short of the aspirations set forth by other BRICS nations. Thus, it appears that Russia continues to prioritize its dominant position as a leading producer of fossil fuels rather than capitalizing on its capacity to become a significant contributor to the global renewable energy landscape.

India's stance mirrors Russia's position with some distinctions. Whereas Russia's tepid enthusiasm for renewable energy arises principally from fossil fuels forming the backbone of governmental income streams, India relies heavily on fossil fuels to sustain robust economic expansion. Presently, approximately 80% of India's energy consumption derives from coal and crude oil, solidifying its profound reliance on non-renewable resources for meeting burgeoning energy needs (IRENA, 2022). Nevertheless, India is actively striving to cultivate renewable energy sources as part of its sustainability-focused strategy. Its declared objective involves multiplying renewable electricity capacity fourfold by 2030 (GoI, 2015); conversely, concurrent plans involve scaling up domestic coal extraction. Given India's already pronounced dependence on fossil fuels, this simultaneous pursuit of opposing paths suggests that gains in renewable energy might ultimately yield insufficient positive outcomes to offset increased fossil fuel usage. Empirical evidence supports this conclusion, evidenced by the rate of growth in non-renewable energy surpassing that of renewable energy in recent years. Projections suggest that ongoing economic ascension amidst increasing population density, accelerated urbanization, and industrialization render it improbable that India will opt for renewable energy over its safer bet—fossil fuels—capable of delivering requisite growth velocities in the near term.

On the other side, we have South Africa which is actively pursuing the policy of coal consumption minimization. We can see that unlike India, who is also heavily dependent on coal, South Africa intends to reduce consumption despite the possibility of facing some short-term issues due to the transition on the background of severe coal dependency. Namely, in 2022 the South African Government announced a long-term Just Energy Transition Partnership with the United States, France, United Kingdom, and Germany. Within this partnership these countries committed to provide to South Africa funding of $8.5 bn during the first phase in order to accelerate decarbonization of the economy. Furthermore, the country has been implementing measures to diversify its energy mix and within the Integrated Resources Plan South Africa is planning to decommission several coal-fired power plants by 2030 and replace them with renewables (ICSC, 2022).

Altogether, the comprehensive analysis conducted herein unequivocally illustrates a trend of escalating divergence in policy objectives and accompanying actions among BRICS nations. This widening difference is attributable to the fundamental differences of their respective national economies, compounded by capacities to facilitate a seamless and incremental transition toward clean energy paradigms without inflicting detrimental repercussions on the current business ecosystem. Moreover, the delicate equilibrium between environmental needs and economic prosperity constrains policymakers' latitude, necessitating careful navigation of tradeoffs. Efforts to minimize disruption to incumbent industries must be balanced against urgent calls for environmental remediation, posing a challenge for governments seeking to reconcile competing imperatives. Consequently, as highlighted earlier, the observed divergence in policy goals and actions reflects deeper structural fault lines embedded within the BRICS grouping, presenting a perplex challenge for cooperative problem-solving and unified climate action.

In terms of future prospects, it is important to acknowledge that although the identified trend raises legitimate concerns regarding the ability of BRICS nations to converge on a unified vision and collaborate effectively, the very existence of the BRICS platform, coupled with the evolving paradigm shift toward multilateralism, presents a promising opportunity. This opportunity allows for the sharing of best practices, mutual learning, and adaptation of successful models, particularly drawing inspiration from Brazil's exceptional track record in renewable energy integration. By leveraging Brazil's proven success stories, other BRICS members could expedite their own transitions toward cleaner energy systems, thereby gradually narrowing the gap in policy coherence and synchronizing their objectives. Such harmonization would enhance the collective influence of BRICS as a credible actor on the global stage, elevating its profile as a thought leader and innovator in addressing climate change challenges through scalable low-emission solutions. Ultimately, for BRICS to transcend its current role as a reactive participant in international forums and evolve into a proactive catalyst for transformative change, it must capitalize on these synergies. By embracing a collaborative mindset, engaging in constructive dialogue, and systematically replicating exemplary cases like Brazil's, BRICS can assert itself as a true architect of multilateral policies and actions, setting precedents for inclusive and sustainable development globally.

4 Closing the Carbon Leakage Gap: Carbon Pricing Mechanisms

4.1 Understanding Carbon Leakage

Transitioning to a green economy, which encompasses measures aimed at mitigating adverse climatic impacts through taxes and pricing schemes, as aforementioned, poses significant policy and financial challenge for government and enterprises. From a governmental perspective, stringent climate policies may pose significant risks to national economic security by negatively impacting the competitive advantage of domestic energy-intensive sectors. Research has demonstrated that rigorous environmental measures often result in substantial increases in energy costs, which subsequently affect industrial productivity and profitability (Jakob, 2021). These heightened expenses create financial burdens for businesses operating within these sectors, leading to potential market disadvantages compared to competitors situated in regions with less restrictive regulatory frameworks. Furthermore, the adverse consequences of elevated energy tariffs extend beyond immediate fiscal concerns, influencing broader macroeconomic indicators such as inflation rates, employment levels, and overall GDP growth trajectories. In this context, developing countries face unique challenges in balancing accelerated industrialization with sustainable resource management, making the formulation of sound policies even more critical. Their pursuit of accelerated growth trajectories demands innovative solutions that reconcile environmental conservation with socio-economic advancement, ensuring inclusive prosperity without compromising ecological integrity. As a consequence, policymakers must carefully balance environmental sustainability objectives against the imperative need to safeguard long-term economic stability and resilience.

From a corporate standpoint, the implementation of rigid environmental policies in certain jurisdictions creates strong incentives for firms to engage in strategic relocations toward areas characterized by comparatively lax regulatory environments (Jakob, 2021). Multinational enterprises based predominantly in advanced economies frequently encounter formidable challenges associated with adhering to demanding ecological norms coupled with escalated operational expenditures. In response, many opt to transfer their most carbon-intensive manufacturing processes overseas, particularly targeting developing nations renowned for their lenient legislative frameworks governing greenhouse gas emissions alongside cost-effective workforce availability and enticing fiscal advantages. Illustrative examples include major multinational enterprises shifting substantial carbon-intensive portions of their production chains to rapidly growing economies like China, India, or Brazil. Historically, such destinations provided not only reduced compliance burdens but also additional stimuli including subsidized utilities, preferential land leases, and expedited administrative procedures aimed explicitly at attracting foreign direct investment. However, this migratory corporate behavior perpetuates what economists refer to as carbon leakage, wherein emitters merely shift their footprint geographically rather than genuinely reducing their total contribution to atmospheric pollution.

Consequentially, the phenomenon of carbon leakage resulted in a consistent decrease in carbon dioxide emissions within industrialized nations, particularly the United States and European Union. Paralleling this trend, there was a concurrent increase in emissions from rapidly industrializing developing economies such as aforementioned China and India. This trend is evident from historical data displayed in Fig. 2. Indeed, throughout much of the early to mid-twentieth century, the bulk of global carbon dioxide emissions originated primarily within European countries and the United States. As far back as 1900, these regions collectively contributed over ninety percent of total annual greenhouse gas (GHG) outputs worldwide. Even up until 1950, their combined share still exceeded eighty-five percent annually. However, beginning with the latter part of the twentieth century, there was an observable shift in emission patterns. This period witnessed substantial increases in fossil fuel consumption and industrial activity outside Western nations, especially across Asian territories such as India but predominantly led by rapid economic growth in China. Consequently, today's figures indicate that while America and Europe continue to contribute substantially to overall emissions, together they no longer dominate as heavily, representing less than one third of current emissions globally.

Aforementioned shifts occurred partially as an outcome of carbon-intensive industries relocating their operations to regions with less stringent environmental regulations. This realignment had significant implications for global emissions patterns, highlighting the inadequacy of isolated national-level climate policies in addressing systemic environmental challenges. Ultimately, this redistribution resulted in elevated global emissions

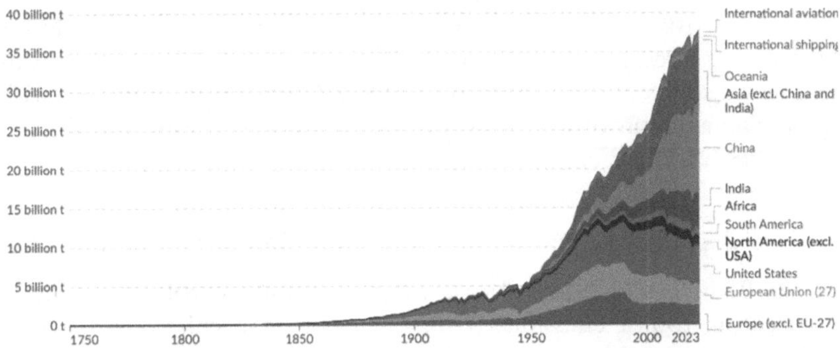

Fig. 2 Annual CO_2 emissions by world region (1750–2023) (*Source* Global Carbon Budget [2024])

attributable to heightened emission intensity within recipient countries (Meng et al., 2023).

Addressing this multifaceted challenge requires concerted cross-border collaboration among all key stakeholders—governments, industry leaders, civil society organizations—to establish equitable yet enforceable guidelines capable of curbing carbon leakage phenomena while promoting environmentally responsible practices across borders. A 2021 study by Henderson and Verma substantiates the need for a cross-border collaboration by presenting evidence that when only a limited number of nations implement a carbon tax, approximately half of their achieved emission cuts are counterbalanced by increased emissions in other non-participating countries. The degree of carbon leakage decreases when more countries join in enforcing such taxes; although higher tax levels lead to greater overall reductions in global emissions, they simultaneously trigger higher rates of emissions displacement, thereby diminishing the effectiveness of high tax rates in scenarios with minimal international participation.

Although the practice of importing carbon emissions to a certain degree allowed developing countries, including BRICS, to accumulate valuable knowledge, know-how, enhance technological capabilities due to the spillover effects of foreign investment in manufacturing capacities, diversify their economies and to overall increase their competitive advantages, they are now acknowledging the need to address this issue by taking into consideration that simply allowing foreign companies to move carbon emissions abroad does not solve global climate challenges. Out of necessity, all BRICS members undertake some sort of action to address this issue. As a result, the trend of introducing carbon pricing mechanisms such as emissions trading schemes (ETS) and carbon tax combined with the implementation of tighter regulation for both domestic and foreign companies has emerged.

4.2 *Carbon Pricing Mechanisms in BRICS Countries*

Within BRICS, China and Russia assume prominent roles as pioneers in executing robust emissions trading schemes. Particularly noteworthy is China's accomplishment since 2021, wherein it successfully instituted a comprehensive carbon pricing mechanism (Chinese national carbon trading scheme) known as the Chinese carbon market, which imposes a cost on greenhouse gas emissions (IRENA, 2022). Under this program, companies receive an allocation of complimentary Certified Emission Allowances (CEAs). In case their actual emissions surpass their allocated quota within a specific compliance period, they need to purchase additional allowances through the market to offset the excess. Conversely, if their emissions fall below the allotted amount, they have the option to sell any remaining CEAs. The distribution of these allowances is determined based on sector-specific carbon intensity standards established by the government, rather than absolute emission volumes, with these benchmarks being progressively lowered over time. Companies are required to provide essential metrics on a monthly basis and disclose annual emission reports.

Since its establishment, this initiative has evolved into the most extensive global emissions trading framework, encompassing approximately 5.1 billion metric tons of CO_2 equivalents, accounting for roughly 40% of China's overall emissions (Reuters, 2024). Building upon this foundational step, Chinese policymakers have additionally signaled intent to augment current regulatory structures by introducing a differentiated industrial carbon levy targeting entities falling beyond the scope of the established ETS. In March 2025, China's Ministry of Ecology and Environment announced plans to broaden the scope of the national Emissions Trading Scheme (ETS), adding cement, steel, and aluminum sectors alongside the existing power sector. Approved by the State Council after public consultations, the move will incorporate 1,500 new companies into the ETS by the end of 2025, raising the total CO2-equivalent emissions under regulation by 3 billion tons—approximately 5% of worldwide emissions (ICAP, 2025).

Russia is likewise evaluating prospective courses of actions to fully introduce a similar scheme. In 2021 the Russian Ministry of Economic Development and the regional government approved a roadmap for implementing an experiment to establish special regulation of greenhouse gas emissions in the Sakhalin region and achieve carbon neutrality by 2025 (IAC RF, 2021). Among the stipulated measures enumerated within this blueprint is the initiation of a trial-stage carbon exchange platform, contingent upon successful demonstration, with aspirations to extend its application across other regions. This roadmap, combined with the Federal Law No. 296-FZ on the Restriction of Greenhouse Gas Emissions dated 2 July 2021, represent preliminary steps aimed at introducing a cohesive framework of carbon regulation in Russia.

Brazil and India, albeit at a slower pace compared to other nations, are nonetheless demonstrating notable advancements within their respective jurisdictions concerning environmental sustainability initiatives. Specifically, as pertinent to India's legislative developments, it was observed that during July 2022, the Lok Sabha—the lower chamber of its bicameral parliamentary system—endorsed an amendment bill to the pre-existing "Energy Conservation Act," originally enacted in 2001 (PRS India, 2022). The salient feature of this legislation lies in its establishment of a framework for fostering a domestically regulated carbon trading market. Additionally, this measure grants explicit authority to relevant governmental agencies for issuing certifications known as carbon credits, which act as quantifiable representations of reduced greenhouse gas emissions achieved by entities participating in such schemes. Subsequently, in December 2022, further endorsement of said amendments transpired when they were ratified by the Rajya Sabha, the upper house of India's parliament. As these provisions stand today, ongoing efforts continue toward operationalizing this novel regulatory mechanism aimed at mitigating climate change impacts through incentivized emission reductions.

Turning attention now toward Brazil, analogous endeavors are being pursued under the auspices of both the national government and key institutions involved in energy policy formulation. On December 21, 2023, the Brazilian House of Representatives gave its approval to an updated version of Bill 2.148/15, designed to establish a regulated carbon trading system known as the Brazilian Greenhouse Gas Emission Trading System

(SBCE). On December 11, 2024 President Luiz Inácio Lula da Silva signed the Law 15.042 which officially established SBCE (GoB, 2024). Analogous to China's and the European Union's ETS, the SBCE establishes a GHG cap-and-trade system based on several types of assets including carbon credit and emissions quota (ICAP, 2024). However, unlike other ETS, the SBCE does not focus on particular industrial sectors but imposes regulation on all of them with the only exception being agriculture. It is expected that the SBCE will become fully operational after the six-year implementation period and will cut Brazil's GHG emissions by at least 59% by 2035 compared to the 2005 level (GoB, 2024). South Africa stands uniquely among the members of the BRICS coalition as the sole nation that neither possesses nor envisions establishing a nationwide ETS within the foreseeable future. Authoritative sources within the country contend that due to the highly concentrated nature of GHG emissions originating predominantly from a restricted subset of enterprises operating within its energy sector, the implementation of an ETS would currently be ill-suited to addressing prevailing challenges effectively (SARS, 2023). This perspective aligns with empirical data highlighting substantial disparities in emission contributions across industrial sectors, thereby necessitating alternative regulatory approaches tailored specifically to local conditions.

In lieu of adopting an ETS, South African policymakers opted instead for enacting the Carbon Tax Act—a comprehensive fiscal instrument designed explicitly to penalize excessive generation of harmful pollutants while simultaneously encouraging ecological responsibility. For instance, firms are able to meet part of their tax liabilities using certified carbon credits. By imposing financial penalties proportional to emitted volumes of GHGs, the taxation regime serves dual purposes: firstly, it creates economic disincentives discouraging continued reliance upon fossil fuel-intensive production methods; secondly, it promotes investment into research and deployment of environmentally friendly alternatives capable of minimizing adverse environmental consequences associated with traditional modes of operation (SARS, 2023). Through strategic utilization of revenue generated via carbon taxes, public funds can then be redirected toward subsidizing green innovations, supporting infrastructure upgrades aligned with low-carbon pathways, and advancing broader societal goals centered around sustainable development principles. Consequently, despite lacking formal mechanisms typical of many developed economies, South Africa demonstrates proactive engagement in combating global warming through targeted interventions geared toward achieving long-term resilience against climate-related risks. Also, it stands to mention that despite the lack of national ETS, South Africa has the JSE Ventures Voluntary Carbon Market launched in 2023 by the Johannesburg Stock Exchange in partnership with US-based environmental markets firm Xpansiv. This carbon market enables local entities to purchase or trade carbon credits and renewable energy certificates which are recorded in either domestic or international registries (JSE, 2023). Beyond employing carbon pricing instruments, albeit to varying degrees, each constituent member state of the BRICS alliance actively pursues complementary strategies aimed at curtailing both direct carbon imports and aggregate emissions levels nationwide. Such efforts manifest primarily through sustained enhancement and refinement of existing regulatory frameworks governing industrial activities, transportation networks, agricultural practices, and urban planning policies. Collectively, these interconnected measures seek to foster innovation-driven solutions, accelerate widespread integration of renewable energy resources, and encourage universal adoption of advanced energy efficiency protocols across all major socioeconomic domains.

Overall, we can conclude that as a result of this emerging trend BRICS countries are gradually becoming the front liners in the efforts to create a global level playing field for industries and encourage both domestic and foreign companies to reduce their emissions. By internalizing the cost of carbon, these measures reduce the incentive for companies to relocate and exploit lower carbon prices or the lack of them. Albeit premature to assess definitive outcomes at present, it is evident that this emergent trajectory holds considerable promise for diminishing deleterious effects and averting future occurrences of the carbon leakage phenomena. Moreover, the accumulated expertise and insights garnered by BRICS nations throughout the execution phase of these systemic transformations position them advantageously to disseminate valuable lessons learned, either bilaterally or multilaterally, assisting fellow developing states confronting analogous challenges. Consequently, such collaborative undertakings serve to consolidate their standing as influential actors within the broader context of global climate governance dynamics.

5 Conclusion

In conclusion, the issue of addressing climate challenges not just within BRICS but globally is getting increasingly complex given that nations must focus on crafting targeted interventions that do not undermine both short-term competitiveness and long-term resilience and facilitate global convergence through balanced approaches.

Overall this chapter analyzed three key trends of addressing climate challenges within BRICS countries:

1. Transition from cooperation based on bilateralism to one based on multilateralism;
2. Divergence of policy goals and actions on the background of significant levels of disparity;
3. Introduction of carbon pricing mechanisms needed to narrow the "carbon leakage" gap, minimize the negative effects and prevent this phenomenon in the future.

Findings indicate that the first and the third trends are positive, while the second one might undermine future prospects in the scope of addressing climate challenges within BRICS given that entities that have different goals and vision on some issue cannot achieve productive and effective cooperation. Therefore, the prospect of effectively addressing climate challenges within BRICS, while at the same time emerging as a vigorous global platform for the formation of a common vision on climate challenges policy, will depend solely on the ability to achieve convergence, align goals and policy actions, which will require a comprehensive and multilateral approach. However, we must note that there are some limitations to this research. The main limitation is the fact that this chapter is devoted only to key areas of BRICS countries' policies in order to highlight the most important trends and does not delve into the specific climate challenges each BRICS country faces. To provide further understanding there is a need to conduct a targeted analysis of each country separately by taking into consideration the national economy's demands and country-unique climate and sustainable growth-related factors. Nonetheless, it can be expected that the provided insight will serve as a starting point for further development of this area of research on trends and prospects of addressing climate challenges within the BRICS group.

References

BRICS Partnership on New Industrial Revolution. (2023). *Joint statement of the ninth (9th) BRICS Environment Ministers Meeting* (30 June 2023). Accessed 15 July 2023. https://www.bricspic.org/En/Pages/Home/NewsDetail.aspx?rowId=509

Centre for Research on Energy and Clean Air. (2023). *Briefing: China permits two new coal power plants per week in 2022.* Accessed 3 August 2023. https://energyandcleanair.org/publication/china-permits-two-new-coal-power-plants-per-week-in-2022/

Global Carbon Budget. (2024). *Annual CO_2 emissions by world region* (1 August 2024). Accessed 12 March 2025. https://explore.globalcarbonbudgetdata.org/

Government of Brazil. (2024). *Press release: President Lula signs law creating regulated carbon market in Brazil* (12 December 2024). Accessed 16 May 2025. https://www.gov.br/planalto/en/latest-news/2024/12/president-lula-signs-law-creating-regulated-carbon-market-in-brazil

Government of India. (2015). *Report on India's renewable electricity roadmap 2030: Toward accelerated renewable electricity deployment*. Accessed on 15 March 2023 from https://www.niti.gov.in/sites/default/files/2023-03/ReportonIndiaRenewableElectricityRoadmap2030.pdf

Henderson, B., & Verma, M. (2021). Global assessment of the carbon leakage implications of carbon taxes on agricultural emissions. *OECD Food, Agriculture and Fisheries Papers, No. 170*, OECD Publishing, Paris.

IAC RF: Russian International Affairs Council. (2021). *Report on emissions trading systems: Trajectories for development* (25 March 2021). Accessed 2 August 2023. https://russiancouncil.ru/en/analytics-and-comments/comments/emissions-trading-systems-trajectories-for-development/

International Carbon Action Partnership. (2025). *China officially expands national ETS to cement, steel and aluminum sectors* (10 April 2025). Accessed 16 May 2025. https://icapcarbonaction.com/en/news/china-officially-expands-national-ets-cement-steel-and-aluminum-sectors

International Carbon Action Partnership. (2024). *Brazilian greenhouse gas emissions trading system* (22 December 2024). Accessed 16 May 2025. https://icapcarbonaction.com/en/ets/brazilian-greenhouse-gas-emissions-trading-system

International Centre for Sustainable Carbon. (2022). *South Africa: Coal usage: Opportunities and challenges in 2022 and beyond*. Accessed on 5 August 2023 from https://www.sustainable-carbon.org/south-africa-coal-usage-opportunities-and-challenges-in-2022-and-beyond/

International Energy Agency. (2020). *Report on China's emissions trading scheme: Designing efficient allowance allocation*. Accessed 1 August 2023. https://www.iea.org/reports/chinas-emissions-trading-scheme

International Energy Agency. (2021). *India's energy outlook*. Accessed 1 August 2023. https://www.iea.org/reports/india-energy-outlook-2021
International Renewable Energy Agency. (2022). *Energy profiles*. Accessed 1 August 2023. https://www.irena.org/Data/Energy-Profiles
Jakob, M. (2021). Why carbon leakage matters and what can be done against it. *One Earth, 4*(5), 609–614.
Johannesburg Stock Exchange. (2023). *JSE collaborates with Xpansiv to launch voluntary carbon market to advance South Africa's carbon credit capabilities* (12 October 2023). Accessed 15 May 2025. https://www.jse.co.za/news/news/jse-collaborates-xpansiv-launch-voluntary-carbon-market-advance-south-africas-carbon
Kirpizli, G. (2022). Through the lenses of morality and responsibility: BRICS, climate change and sustainable development. *Uluslararası İlişkiler / International Relations, 19*(75), 65–82.
Meng, J., Huo, J., & Zhang, Z., et al. (2023). The narrowing gap in developed and developing country emission intensities reduces global trade's carbon leakage. *Nature Communications, 14*, 3775.
Ministry of Finance of the Russian Federation. (2023). *Information on additional oil and gas revenues of the federal budget* (1 May 2023). Accessed 25 May 2023. https://minfin.gov.ru/
Ministry of Foreign Affairs of Brazil. (2023). *Brazil-China Joint Statement on combating climate change* (14 April 2023). Accessed 25 May 2023. https://www.gov.br/mre/en/contact-us/press-area/press-releases/brazil-china-joint-statement-on-combating-climate-change
People's Daily. (2023). *China, Russia continue expanding cooperation in green development*. (18 August 2021). Accessed 14 May 2023. http://en.people.cn/n3/2023/0322/c90000-10225763.html
PRS Legislation Research India. (2022). *The Energy Conservation (Amendment) Bill*. Accessed 31 July 2023. https://prsindia.org/billtrack/the-energy-conservation-amendment-bill-2022
Reich, A. (2009). Bilateralism versus multilateralism in international economic law: Applying the principle of subsidiarity. *University of Toronto Law Journal, 60*(2).
Reuters. (2024). *What's China's carbon market and how does it work?* (12 September 2024). Accessed 16 May 2025. https://www.reuters.com/sustainability/whats-chinas-carbon-market-how-does-it-work-2024-09-12/
Ruggerio, C. A. (2021). Sustainability and sustainable development: A review of principles and definitions. *Science of the Total Environment, 786*.
Ruggie, J. G. (1993). *Multilateralism matters: The theory and praxis n an international form*. Columbia University Press.
Ruppel, O. C., & Ruppel-Schlichting, K. (2013). The BRICS partnership: Development and climate change policy from an African Perspective. In *Climate*

change: International law and global governance volume II: Policy, diplomacy and governance in a changing environment (pp. 549–569). Baden-Baden.

SC PRC: The State Council of the People's Republic of China. (2021). *BRICS space agencies in deal to share remote sensing satellite data* (18 August 2021). Accessed 14 May 2023. https://english.www.gov.cn/statecouncil/ministries/202108/18/content_WS611d0e95c6d0df57f98dea50.html

South African Revenue Service. (2021). *Carbon tax*. Accessed 2 August 2023. https://www.sars.gov.za/customs-and-excise/excise/environmental-levy-products/carbon-tax/

Tian, Xu., & Sarkis, J. et al. (2020). Examining the role of BRICS countries at the global economic and environmental resources nexus. *Journal of Environmental Management, 262.*

Weber, S. (1991). *Multilateralism in NATO: Shaping the postwar balance of power, 1945–1961.* University of California at Berkeley.

World Economic Forum. (2022). *Energy transition investment: Top 10 Countries in 2022* (15 February 2022). Accessed 25 July 2023. https://www.weforum.org/stories/2022/02/ranked-10-countries-energy-transition-investment/

CHAPTER 10

How Effective Are Sustainable Marketing Strategies in the South African Consumer Market? A Systematic Review

Thembeka Goodness Zondi, Paballo Patience Ntobaki, and Nkululeko PraiseGod Zungu

1 Introduction

Sustainability has become a central concern in global business and marketing discourse, particularly as climate change, ecological degradation, and resource constraints continue to challenge traditional consumption patterns. In South Africa, these global crises converge with local issues such as inequality, unemployment, and urban expansion, creating a complex socio-economic landscape that calls for sustainable, inclusive, and culturally responsive solutions (Majeed et al., 2022; Tan et al., 2022). Sustainable marketing has emerged as a promising tool to influence environmentally and socially responsible consumption behaviours. It comprises practices such as green product development, eco-labelling,

T. G. Zondi · P. P. Ntobaki · N. P. Zungu (✉)
Department of Marketing and Retail, Durban University of Technology, Durban, South Africa
e-mail: NkululekoZ1@dut.ac.za

N. Tshikovhi et al. (eds.), *The BRICS Imperative*,
https://doi.org/10.1007/978-3-032-05932-1_10

ethical branding, and transparent communication about environmental impacts (Peterson et al., 2021; Rehman et al., 2021). In the fast-moving consumer goods (FMCG) sector, which plays a pivotal role in daily consumption routines, such strategies are especially critical due to their potential to shape large-scale behavioural shifts across diverse consumer segments (Haider et al., 2022).

Recent evidence suggests that sustainable marketing can significantly improve consumer attitudes, brand trust, and purchase intentions particularly when it is supported by credible environmental messaging and culturally relevant branding (Majeed et al., 2022). However, in South Africa, the practical implementation and impact of these strategies remain uneven. While middle- and high-income consumers are more likely to engage in green purchasing behaviours, many lower-income consumers continue to prioritise price and availability over sustainability due to economic constraints and infrastructural limitations (Tan et al., 2022). Trust and transparency have also emerged as critical factors. Peterson et al. (2021) note that consumer scepticism is rising due to widespread greenwashing and misleading corporate claims, which undermine the credibility of sustainability marketing efforts. Moreover, structural barriers such as limited access to sustainable goods, weak enforcement of labelling standards, and a general lack of consumer education further exacerbate the gap between sustainability awareness and action (Haider et al., 2022).

The need for contextualised, evidence-based marketing strategies is therefore urgent particularly in developing economies like South Africa, where socio-economic disparities affect how sustainability is perceived, prioritised, and enacted. As Rehman et al. (2021) argue, sustainable performance is only possible when marketing is strategically integrated with environmental control systems, innovation, and community engagement. In this context, the current study investigates the role of sustainable marketing in influencing South African consumer behaviour, with a specific focus on the FMCG sector. It seeks to understand how consumers perceive, interpret, and respond to sustainable marketing strategies, and what factors enhance or inhibit their engagement with such efforts. The study is guided by the following research questions:

1. What sustainable marketing strategies are currently being implemented in the South African consumer market?
2. How do South African consumers perceive and respond to these strategies?

3. What factors influence the success or failure of these strategies?

2 Literature Review

2.1 *Conceptualising Sustainability and its Marketing Relevance*

Sustainability is widely understood as the capacity of current generations to meet their needs without compromising the ability of future generations to meet theirs (Mensah, 2019). Within this framework, sustainable marketing has emerged as a strategic response to the ecological and social pressures associated with rapid industrialisation, climate change, and irresponsible consumption. In fast-moving consumer goods (FMCG) markets characterised by high consumption frequency and short product life cycles sustainable marketing seeks to integrate environmental responsibility into everyday consumer choices (Veselova, 2022). Sustainability marketing, a prominent subset of sustainable marketing, focuses on promoting environmentally friendly products using tools such as eco-labelling, sustainable packaging, and ethical branding. These strategies contribute to long-term pollution reduction and support national and international environmental agendas (Rahman et al., 2022). However, the majority of scholarship in this domain originates from developed economies, resulting in a knowledge gap about how sustainable marketing operates in emerging markets like South Africa (Reddy et al., 2023). Addressing this gap is critical, as sustainable marketing in developing contexts must adapt to distinct economic, social, and infrastructural conditions.

2.2 *Determinants of Sustainable Consumer Behaviour*

Green consumerism reflects the evolving role of the consumer in advancing sustainability. Kumar et al. (2021) note that sustainable consumers increasingly seek products that align with their ethical, environmental, and social values. These consumers evaluate the entire product lifecycle, from raw material sourcing to post-use disposal. Yet, while sustainable consumption has gained traction globally, its adoption in Africa remains slow due to structural and behavioural challenges (Reddy et al., 2023). Consumers' environmental awareness is growing, but this often does not translate into actual purchase behaviour. Xie et al. (2022) found that consumers may express pro-environmental attitudes but remain inactive due to behavioural inertia, lack of affordability, or limited access. In South Africa, these barriers are exacerbated by income disparities and weak infrastructure (Reddy et al., 2023). Additionally, trust in sustainability claims remains low, especially when companies engage in greenwashing exaggerating or falsifying their environmental credentials (Gelderman et al., 2021). Cultural and linguistic diversity also shapes green consumer behaviour. According to Rahman et al. (2022), sustainable marketing campaigns that do not consider local languages or cultural values fail to resonate with diverse consumer segments. This highlights the need for localised, inclusive marketing efforts that go beyond Western sustainability narratives.

2.3 Strategic Approaches to Sustainable Marketing in FMCG

The literature points to a variety of sustainable marketing strategies currently implemented within the South African consumer market. These include green product design, ethical branding, circular packaging solutions, and CSR-driven campaigns. Rahman et al. (2022) and Kumar et al. (2021) argue that authentic, transparent, and holistic strategies are more likely to influence consumer perception and loyalty. Effective campaigns often communicate not just the environmental benefits of a product but also social value such as job creation, community empowerment, or affordability. For instance, ethical branding strategies, which position companies as socially responsible entities, can foster strong emotional connections with consumers. In the South African context, where trust in both public and corporate institutions is limited, such credibility must be earned through consistency and accountability (Gelderman et al., 2021). Technological innovation is another pillar of sustainable marketing. The use of mobile apps, QR code traceability, and digital storytelling tools enables companies to communicate sustainability credentials more transparently. These tools also serve educational purposes by raising environmental awareness (Xie et al., 2022).

2.4 Barriers to Effective Sustainable Marketing

Despite growing corporate interest in sustainability, significant barriers impede the widespread adoption of green marketing practices. Economic constraints remain the most prominent. Reddy et al. (2023) and Rahman et al. (2022) report that sustainable products are often priced higher than conventional alternatives, making them inaccessible to low-income groups. In South Africa, where a substantial portion of the population faces financial hardship, the higher cost of sustainable products discourages green consumption. Another critical barrier is infrastructural underdevelopment. Many rural and informal communities lack access to recycling facilities, eco-friendly retail outlets, and public environmental services. According to Kumar et al. (2021), this limited infrastructure undermines consumer participation, even among those who are environmentally conscious. A further challenge is the lack of consumer education. Without understanding what sustainability entails or how to identify green products, consumers are less likely to make informed choices. Xie et al. (2022) emphasise that environmental literacy is a key enabler of green consumer behaviour. This calls for investments in educational campaigns, particularly those that are culturally sensitive and linguistically accessible.

2.5 *Towards a Contextualised Sustainable Marketing Framework*

In response to these challenges, researchers call for a contextualised sustainable marketing framework tailored to the realities of South African consumers. Such a framework would integrate economic inclusivity, cultural relevance, and infrastructural development. As Gelderman et al. (2021) note, credibility and transparency are non-negotiable in building consumer trust. Meanwhile, Rahman et al. (2022) suggest coupling sustainable marketing with social development goals to enhance resonance and participation (Fig. 1).

The Contextualised Sustainable Marketing Framework depicted in the diagram provides a strategic lens through which sustainable marketing in South Africa can be understood and implemented. It calls for a localised approach that goes beyond generic global narratives to address context-specific challenges such as inequality, infrastructural limitations, and consumer mistrust. South Africa's high levels of income disparity necessitate inclusive marketing strategies that offer value across income brackets. Economic inclusivity in sustainable marketing includes designing affordable eco-friendly products, promoting inclusive supply chains, and supporting local entrepreneurship to enhance economic resilience. Sustainable brands that create jobs and uplift communities are more likely to win local support.

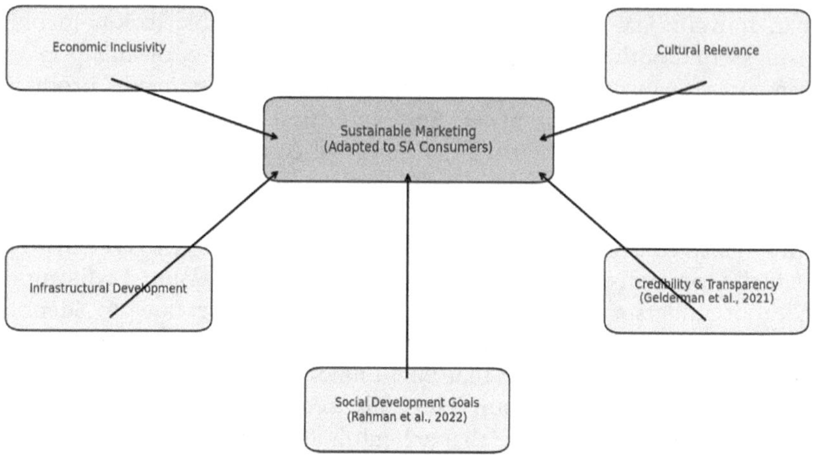

Fig. 1 Contextualised sustainable marketing framework

Marketing strategies must reflect and respect South Africa's cultural and linguistic diversity. Culturally relevant marketing builds deeper resonance and participation (Rahman et al., 2022). As supported by Bernard et al. (2024), sustainability narratives must be framed in ways that connect with local traditions, languages, and values to promote long-term behavioural shifts. Practical implementation of sustainable marketing depends on infrastructure such as access to recycling systems, transport, and electricity which remains unevenly distributed in South Africa. Sustainable marketing frameworks must therefore consider infrastructural constraints and align with developmental efforts. This also includes adapting product designs and messaging to accommodate informal markets and rural consumers. Credibility and transparency are foundational to consumer trust. Gelderman et al. (2021) emphasise that clear, honest communication and traceable sustainability claims are essential in mitigating scepticism. Without these, consumers may perceive sustainable marketing as greenwashing. Therefore, third-party certifications, transparent reporting, and inclusive stakeholder engagement become critical tools.

Rahman et al. (2022) advocate for aligning sustainable marketing efforts with social development agendas such as the SDGs. In the South African context, this includes SDG 12 (responsible consumption and production) and SDG 3 (good health and well-being). Campaigns that raise awareness about healthy, sustainable living while also addressing social inequalities are more impactful and relevant. The framework's strength lies in its integration. Economic inclusivity facilitates broader access; cultural relevance enhances emotional engagement; infrastructure enables action; transparency builds trust; and SDG alignment grounds marketing in social purpose. Together, these create a dynamic, actionable framework suited for sustainable transformation in the South African market.

3 Methodology

This study adopts a systematic literature review (SLR) to evaluate the effectiveness of sustainable marketing strategies in the South African consumer market. A systematic review offers a transparent and replicable method of synthesising existing empirical evidence to answer focused research questions. The Preferred Reporting Items for Systematic Reviews and Meta-Analyses (PRISMA) framework was applied to guide article selection and thematic synthesis. A comprehensive literature search was conducted using academic databases such as Google Scholar, Elsevier, Emerald Insight, MDPI, Springer Nature, and Taylor and Francis. Grey literature, including NGO and industry reports, was also reviewed to ensure contextual relevance. The following search terms were used in various combinations with Boolean operators (AND, OR): "sustainable marketing," "FMCG," "consumer behaviour," "greenwashing," "eco-friendly products," and "South Africa."

3.1 Inclusion and Exclusion Criteria

To ensure depth and accuracy, studies were filtered using predefined criteria (Table 1).

3.2 Screening Process and PRISMA Summary

An initial 278 records were identified. After removing 74 duplicates, 204 titles and abstracts were screened, excluding 124 for irrelevance. Of the 80 full-text studies reviewed, 49 were excluded, resulting in 31 eligible studies used for the final synthesis. These are summarised Table 2.

Table 1 Inclusion and exclusion criteria

Inclusion criteria	*Exclusion criteria*
Published from 2019 onwards	Published before 2019
Empirical studies (quantitative, qualitative, or mixed methods)	Conceptual or theoretical papers only
Peer-reviewed journal articles	Non-English texts
Focused on sustainable marketing or green consumer behaviour	Not aligned with sustainable consumption or marketing
South African context or comparable emerging markets	Lacked primary evidence or relevance

Table 2 PRISMA summary

Stage	*Number of studies*
Records identified (database + grey)	278
Duplicates removed	74
Titles and abstracts screened	204
Excluded at screening	124
Full texts assessed	80
Excluded at full-text stage	49
Final included studies	**31**

3.3 Data Extraction and Analysis

A data extraction matrix was developed to capture core variables including author(s), year, objectives, methodology, findings, and relevance to the South African FMCG context. Thematic analysis was then conducted using a comparative matrix approach. This allowed for synthesis under major themes identified in the following section, such as:

- Sustainable marketing mix strategies
- Consumer attitudes versus behaviours
- Economic constraints
- Trust and greenwashing
- Environmental awareness
- Cultural localisation
- Infrastructure and access barriers

Each theme was supported by empirical findings from peer-reviewed studies published from 2020 onwards, ensuring contemporary and policy-relevant observations.

4 Presentation of Findings and Discussion

4.1 *Findings*

The findings of this study reveal that while sustainable marketing strategies are increasingly being deployed in South Africa, their success is mediated by a complex interplay of economic, cultural, informational, and infrastructural factors. Strategies must go beyond surface-level green claims and deeply engage with the structural realities of the South African consumer market. Authenticity, affordability, local relevance, education, and accessibility emerge as critical elements of any successful sustainable marketing intervention. This is reflected in the Table 3.

4.2 *Discussion of Results*

4.2.1 Theme 1: Sustainable Marketing Mix Strategies

Sustainable marketing mix strategies have evolved into comprehensive frameworks aimed at embedding environmental and social responsibility across the entire marketing process. According to Gelderman et al. (2021), applying these strategies consistently increases green consumer satisfaction and boosts perceptions of brand authenticity. This aligns with Rahman et al. (2022), who found that integrating green marketing with broader sustainability development goals fosters deeper consumer trust and long-term loyalty, ultimately improving environmental and financial performance. In the South African context, where trust and value are central to consumer decision-making, these practices become especially relevant. Kumar et al. (2021) argue that true sustainable marketing requires strategic integration rather than superficial tactics. Green strategies must permeate the entire value chain, from sourcing to post-consumer disposal, to avoid the risks associated with tokenism or greenwashing. Product innovation that highlights lifecycle impact (e.g. biodegradable packaging, reduced emissions) and communicates these benefits transparently is increasingly important in building consumer trust and stimulating green purchasing. In South Africa, where consumers are becoming more discerning, genuine value-driven green messaging is

Table 3 Summary of key findings

Theme	*Area of focus*	*Conclusion drawn*	*References*
Sustainable Marketing mix strategies	Use of product innovation, eco-labelling, and green brand positioning	Sustainable strategies increase consumer satisfaction and brand value when authentic and consistent	Gelderman et al. (2021); Rahman et al. (2022); Kumar et al. (2021)
Positive attitudes but low adoption	Consumer interest in eco-friendly FMCG but low follow-through in purchasing	Environmental concern does not always translate to purchase due to financial and behavioural gaps	Reddy et al. (2023); Xie et al. (2022)
Economic constraints	Affordability of sustainable goods among lower-income groups	Price remains the biggest barrier for green product adoption among South African consumers	Reddy et al. (2023)
Trust and greenwashing	Impact of false environmental claims on consumer trust	Perceived dishonesty weakens brand equity and reduces sustainable product uptake	Rahman et al. (2022); Gelderman et al. (2021)
Environmental knowledge and awareness	Effect of environmental education on green consumer behaviour	Knowledge enhances the impact of green marketing and drives behavioural change	Xie et al. (2022); Kumar et al. (2021)
Cultural and linguistic fit	Localisation of marketing messages and cultural sensitivity	Marketing efforts must align with diverse consumer values to be effective	Rahman et al. (2022); Xie et al. (2022)
Infrastructure and access issues	Availability of eco-friendly products and supporting systems (e.g. recycling)	Green marketing success is undermined by poor infrastructure and supply chain constraints	Reddy et al. (2023); Kumar et al. (2021)

critical for sustainable product uptake. Moreover, companies that apply eco-labelling, ethical branding, and carbon footprint disclosures are better positioned to influence green purchase behaviour. These practices not only build brand equity but also act as educational tools for consumers. Brands like Woolworths and Faithful to Nature in South Africa have leveraged these approaches effectively, positioning themselves as leaders in ethical consumption.

4.2.2 *Theme 2: Positive Attitudes but Low Adoption*

Despite the rise in environmental awareness, consumer behaviour does not always reflect sustainability values. Reddy et al. (2023) found a significant disparity between green consumer attitudes and actual purchase decisions in South Africa. Consumers express concern for environmental issues and indicate preferences for sustainable FMCG products, yet these intentions often do not result in sales. This behavioural inconsistency is compounded by economic pressures and ingrained shopping habits, creating a disjunction between ideals and actions. Xie et al. (2022) explain that behavioural inertia, often shaped by habitual consumption patterns and perceived inconvenience, inhibits change. Additionally, without immediate and tangible benefits, consumers may be reluctant to switch to green alternatives. In South Africa's developing economy, where short-term survival often trumps long-term ecological goals, marketing strategies must address this intention-action gap by linking green choices to direct consumer benefits such as affordability, health, or community upliftment. This theme suggests a need for incentive-based marketing promotions, loyalty programmes, or subsidies linked to green purchases. Public–private partnerships may play a role in supporting these behavioural nudges at scale.

4.2.3 *Theme 3: Economic Constraints*

Economic inequality remains a major barrier to sustainable consumption in South Africa. Reddy et al. (2023) underscore that sustainable goods are often seen as premium or niche products, unaffordable to a large portion of the population. This economic divide places pressure on marketers to make sustainability accessible to all socio-economic groups. Incorporating affordability into the sustainable marketing mix is not just a moral imperative but a practical one. As Kumar et al. (2021) suggest, sustainable products need to be priced competitively to achieve mass-market adoption. This requires innovation not only in product development but also in cost-effective sourcing, supply chain design, and localised production. For instance, scaling up community-based manufacturing of sustainable goods could offer both economic upliftment and affordability. Furthermore, bundling sustainable benefits (e.g. health, cost-savings from reusables) with core product messages can redefine perceived value. Social entrepreneurs in South Africa have already demonstrated this through low-cost solar lighting and refillable cleaning products targeted at lower-income markets.

4.2.4 *Theme 4: Trust and Greenwashing*

Trust is foundational in shaping consumer responses to sustainable marketing. Rahman et al. (2022) point out that consumers increasingly scrutinise brand claims and are quick to detect inconsistencies. In cases of greenwashing, where companies overstate or falsify their environmental efforts, the damage to consumer trust can be severe and long-lasting. Gelderman et al. (2021) reinforce this by showing that consumer satisfaction with green products is tightly linked to perceived credibility and transparency. South African consumers, already wary due to historical issues of corporate accountability, are particularly sensitive to disingenuous sustainability claims. This calls for brands to invest in third-party certifications, transparent sustainability reporting, and consistent messaging to maintain integrity and rebuild confidence. Transparency measures such as QR codes linking to sustainability reports, visible certifications, and community engagement campaigns are practical steps that can increase authenticity. In this way, marketing becomes not only a communication tool but a platform for accountability.

4.2.5 *Theme 5: Environmental Knowledge and Awareness*

Environmental literacy plays a crucial role in enabling consumers to make informed choices. Xie et al. (2022) found that well-informed consumers are more likely to respond positively to green marketing and exhibit repeat buying behaviour. Kumar et al. (2021) note that educational efforts, such as sustainability workshops, in-store campaigns, or product labelling, significantly contribute to consumer understanding and brand trust. In South Africa, where educational inequality is widespread, targeted educational interventions are essential. Marketers can collaborate with schools, local NGOs, and government programmes to raise awareness about environmental issues and introduce practical sustainability concepts. Such initiatives not only foster behavioural change but also support community development goals. Importantly, environmental education should not be a one-size-fits-all approach. Using culturally appropriate and linguistically accessible materials can increase reach and efficacy. Gamification, storytelling, and community-driven education have shown promise in increasing participation in green behaviour in township and rural contexts.

4.2.6 *Theme 6: Cultural and Linguistic Fit*

Effective green marketing in South Africa must reflect the country's rich cultural and linguistic diversity. Rahman et al. (2022) stress that campaigns lacking local cultural references often fail to engage intended audiences. Similarly, Xie et al. (2022) highlight the importance of language and storytelling in crafting resonant messages that align with community norms and values. Localising content, whether through isiZulu radio spots, Xhosa pamphlets, or township community events, ensures inclusivity and relevance. Cultural sensitivity in visual design, tone, and messaging can enhance trust and foster stronger brand-consumer relationships. In a multicultural society like South Africa, standardised global campaigns often appear tone-deaf; localised marketing presents a strategic advantage. Community co-creation can also enhance message authenticity. Involving local leaders, traditional authorities, and youth ambassadors in green campaigns ensures cultural alignment and trust. This participatory approach not only improves campaign effectiveness but also empowers local ownership of sustainability goals.

4.2.7 *Theme 7: Infrastructure and Access Issues*

Green marketing efforts are constrained by infrastructural challenges in product availability and supporting systems like recycling and distribution. Reddy et al. (2023) reveal that many South African consumers, especially in rural and informal settlements, lack access to eco-friendly goods due to poor supply chains. Even when demand exists, logistical bottlenecks limit market penetration. Kumar et al. (2021) add that successful green marketing must be supported by tangible accessibility. For example, placing recycling bins in public areas, offering green products in township supermarkets, or facilitating mobile refilling stations for household products are all practical solutions. Without such structural support, even the most compelling green campaigns will fall short of producing real change. Further, decentralised retail models, mobile commerce, and green distribution hubs can help extend the reach of sustainable goods. Public infrastructure, such as dedicated recycling zones or public awareness kiosks, complements private sector initiatives, building an enabling environment for sustainable consumption.

5 Conclusion and Recommendations

5.1 Conclusion

This study has critically examined the role of sustainable marketing in influencing consumer behaviour within the South African fast-moving consumer goods (FMCG) sector, a context marked by persistent inequalities, rising environmental pressures, and evolving consumer awareness. Against this backdrop, the study aimed to explore the nature, reception, and efficacy of sustainable marketing strategies implemented in the country, guided by three research questions concerning the strategies employed, consumer perceptions, and factors influencing success or failure. Firstly, the study revealed that a range of sustainable marketing practices are present in the South African FMCG market. These include green packaging innovations, eco-labelling, sustainability-focused corporate social responsibility (CSR) initiatives, and environmentally themed promotional campaigns (Majeed et al., 2022; Rahman et al., 2022). However, the extent of implementation is influenced by firm size and capacity. Large multinational corporations tend to demonstrate greater sophistication in integrating sustainability into their value chains and communication strategies, while smaller domestic brands often face financial, infrastructural, or knowledge constraints that limit their engagement with green marketing (Reddy et al., 2023).

Secondly, the analysis showed that consumer perceptions are not uniform across demographic and socio-economic segments. Environmental consciousness is growing, particularly among urban, educated, and higher-income consumers who are more likely to support brands that align with their values. However, price sensitivity, access constraints, and limited environmental literacy continue to restrict sustainable consumption among lower-income consumers (Tan et al., 2022). Furthermore, the credibility of marketing claims emerged as a critical concern. Consumers frequently express scepticism due to inconsistent labelling practices, superficial environmental claims, or perceptions of greenwashing, which erode trust in sustainability messaging (Peterson et al., 2021; Xie et al., 2022).

Thirdly, the study identified a complex interplay of factors that influence the effectiveness of sustainable marketing. Among these, affordability and accessibility are paramount; sustainable products must be economically viable for broader segments of the population if they are to achieve mass adoption (Veselova, 2022). Credibility and transparency are essential in building long-term brand trust, and strategies rooted in cultural relevance tend to resonate more deeply with local audiences (Gelderman et al., 2021; Rahman et al., 2022). Additionally, the success of sustainable marketing is shaped by broader structural enablers, including policy support, recycling and green logistics infrastructure, and collaborative private—public partnerships that create enabling environments for sustainable consumption (Rehman et al., 2021).

This study confirms that sustainable marketing, while conceptually promising, cannot be effective in South Africa without strategic localisation and inclusivity. It must be recognised that sustainability is not just a marketing trend, but a developmental necessity that must be inclusive of all consumers regardless of income or literacy level. For sustainable marketing to move from rhetoric to reality, it must be implemented in ways that reflect South Africa's cultural diversity, economic complexity, and infrastructural limitations. The study makes a meaningful contribution by demonstrating that contextualised sustainable marketing frameworks, grounded in socio-economic realities and supported by transparent communication, are essential for shifting consumer behaviour and advancing responsible consumption in South Africa. These findings also hold broader relevance for other developing and emerging economies facing similar structural constraints and demographic diversity.

Future research should expand this inquiry by incorporating primary empirical data (e.g. consumer surveys, interviews, or experiments) to test behavioural responses to different green marketing messages and formats. Moreover, comparative studies across African or BRICS economies could provide further insight into how sustainability strategies evolve across different development contexts.

5.2 *Recommendations*

Based on the findings and the conclusion of this study, the following recommendations are made for marketers, policymakers, and stakeholders in South Africa:

5.2.1 Develop Inclusive Pricing and Product Strategies

To improve accessibility, brands should introduce tiered product ranges that offer affordable green alternatives without compromising core sustainability principles. Partnering with local suppliers, cooperatives, and informal retailers could also help to lower costs and enhance reach.

5.2.2 *Invest in Consumer Education and Environmental Literacy*

Brands and public institutions must engage in consumer education campaigns that demystify green products and build understanding of sustainability terms, labels, and benefits. Digital platforms, schools, and community outreach programmes should be utilised to make this knowledge more widespread, especially in under-served areas.

5.2.3 *Ensure Transparency and Combat Greenwashing*

Marketers should adopt clear, honest, and evidence-based communication around their sustainability efforts. Utilising third-party eco-labels, lifecycle data, or digital QR codes can provide traceability and increase trust in sustainability claims. The Consumer Protection Act should also be updated to better regulate misleading green claims.

5.2.4 *Leverage Cultural and Social Norms*

Marketing messages should be crafted to reflect South African values, identities, and traditions, including Ubuntu, community resilience, and environmental stewardship. Local influencers and trusted community figures can serve as ambassadors for green products, enhancing relatability and trust.

5.2.5 *Encourage Innovation in Green Product Design*

Firms should invest in R&D for sustainable packaging, supply chains, and product innovation tailored to the realities of the South African market. This includes biodegradable packaging, refillable systems, and region-specific materials that reduce costs and environmental impact (Roh et al., 2022).

5.2.6 *Monitor and Evaluate Sustainable Marketing Outcomes*

To avoid performative sustainability, brands must monitor the outcomes of their sustainable marketing strategies, using tools such as consumer surveys, sales data, and environmental impact assessments. Continuous improvement should be embedded into strategy cycles.

References

Chou, S.-F., Horng, J.-S., Liu, C.-H., Yu, T.-Y., & Kuo, Y.-T. (2022). Identifying the critical factors for sustainable marketing in the catering: The influence of big data applications, marketing innovation, and technology acceptance model factors. *Journal of Hospitality and Tourism Management, 51*, 11–21. https://doi.org/10.1016/j.jhtm.2022.02.010

Gelderman, C. J., Paulraj, A., & Van Weele, A. J. (2021). Green marketing as an environmental practice: The impact on green satisfaction and green loyalty. *Business Strategy and the Environment, 30*(2), 839–854. https://doi.org/10.1002/bse.2646

Haider, M., Shannon, R., & Moschis, G. P. (2022). Sustainable consumption research and the role of marketing: A review of the literature (1976–2021). *Sustainability, 14*(7), 3999. https://doi.org/10.3390/su14073999

Kumar, S., Dixit, A., Javalgi, R. G., Dass, M., & Kroll, M. (2021). What do we know about business strategy and environmental research? Insights from Business Strategy and the Environment. *Business Strategy and the Environment, 30*(8), 3445–3465. https://doi.org/10.1002/bse.2871

Majeed, M. U., Aslam, S., Murtaza, S. A., Attila, S., & Molnár, E. (2022). Green marketing approaches and their impact on green purchase intentions: Mediating role of green brand image and consumer beliefs towards the environment. *Sustainability, 14*(18), 11703. https://doi.org/10.3390/su141811703

Mensah, J. (2019). Sustainable development: Meaning, history, principles, pillars, and implications for human action: Literature review. *Cogent Social Sciences, 5*(1), 1653531. https://doi.org/10.1080/23311886.2019.1653531

Peterson, M., Minton, E. A., Liu, R. L., & Bartholomew, D. E. (2021). Sustainable marketing and consumer support for sustainable businesses. *Sustainable Production and Consumption, 27*, 157–168. https://doi.org/10.1016/j.spc.2020.10.018

Rahman, M. M., Park, J., & Chi, G. C. (2022). Towards sustainable development: Coupling green marketing strategies and consumer behavior. *Business Strategy and the Environment, 31*(2), 623–636. https://doi.org/10.1002/bse.2906

Reddy, K., Moeng, C., Hlongwane, D., & Mabaso, B. (2023). Consumers' perception on green marketing towards eco-friendly fast-moving consumer goods. *International Journal of Green Marketing and Business Strategy, 2*(1), 12–24.

Rehman, S. U., Bhatti, A., Kraus, S., & Ferreira, J. J. M. (2021). The role of environmental management control systems for ecological sustainability and sustainable performance. *Management Decision, 59*(9), 2217–2237. https://doi.org/10.1108/MD-06-2020-0800

Roh, T., Noh, J., Oh, Y., & Park, K.-S. (2022). Structural relationships of a firm's green strategies for environmental performance: The roles of green supply chain management and green marketing innovation. *Journal of Cleaner Production, 356*, Article 131877. https://doi.org/10.1016/j.jclepro.2022.131877

Tan, Z., Sadiq, B., Bashir, T., Mahmood, H., & Rasool, Y. (2022). Investigating the impact of green marketing components on purchase intention: The mediating role of brand image and brand trust. *Sustainability, 14*(10), 5939. https://doi.org/10.3390/su14105939

Veselova, A. (2022). Sustainable FMCG consumption and marketing in developing countries: Opportunities and constraints. *Journal of Consumer Behaviour in Emerging Markets, 15*(3), 214–228.

Xie, X., Zhang, Y., Lu, L., & Geng, Y. (2022). Green consumer behavior: Influence of environmental awareness and perceived product value. *Sustainability, 14*(13), 6107. https://doi.org/10.3390/su14136107

CHAPTER 11

The Role of Marketing in Reducing Carbon Emissions: Perspectives from the South African Consumer Market

Paballo Patience Ntobaki and Ndivhuho Tshikovhi

1 Introduction

According to Berndt et al. (2021: 5), marketing consists of a number of institutions and procedures for developing, conveying, delivering, and exchanging offerings that are valuable to customers, clients, partners, and society as a whole. In order to achieve organisational objectives, the traditional marketing idea has put a strong emphasis on meeting customer needs and wants. But in a day of limited resources, environmental degradation, and global competitiveness, the societal perspective

P. P. Ntobaki (✉)
Department of Marketing and Retail, Durban University of Technology, Durban, South Africa
e-mail: paballon@dut.ac.za

N. Tshikovhi
Faculty of Management Sciences, Durban University of Technology, Durban, South Africa

N. Tshikovhi et al. (eds.), *The BRICS Imperative*,
https://doi.org/10.1007/978-3-032-05932-1_11

also emphasises how marketing impacts society as a whole. This sociocultural viewpoint asks if meeting client needs is in society's best interests over the long-term. Businesses that adopt the societal orientation make ethical and socially conscious marketing choices. They must take into account everyone impacted by their operations in addition to just the immediate demands of consumers. Marketing should work to maintain and improve societal wellbeing over the long run. As well as environmentalism, consumer rights, ethical concerns, and social responsibility (Strydom, 2016: 4), marketing has also been linked to environmental harm because its activities have shortened product life cycles and increased waste and consumption (Sharma, 2021: 1217). This is due to the fact that humans have always been consumers because they have continuously consumed numerous items to satisfy their requirements, which include survival and self-esteem (Goworek & McGoldrick, 2015: 4).

It has been found that the majority of young people think about environmental issues before making a purchase or hiring a service. The majority of shoppers are, however, perplexed by the design and cost of eco-friendly clothing. Because eco-friendly clothing is more expensive than regular clothing, people are reluctant to choose it. They are also uninformed of the true environmental benefits (Kumar et al., 2021: 2). Climate change and biodiversity loss are two examples of the irreversible environmental repercussions of human activity. Each person's ecological footprint is largely the result of their product consumption. In order to achieve more sustainable lifestyles, it is crucial to produce and consume more ecologically friendly goods. Environmentally friendly or "green" products are currently, however, still a niche market (Ketelsen et al., 2020: 3). There is no way to overstate the connection between environmental degradation and economic growth. In addition to carbon dioxide, other emissions that contribute to environmental pressure include particulate matter. Over time, these emissions have an impact on commerce and per capita income. In order to enable sustainable development, strong policy, and regulatory institutions are needed to boost economic growth while lowering harmful emissions. (Abbasi et al., 2021: 1441) In the long run, GDP, trade, and urbanisation positively correlate with CO2 emissions; this relationship may differ by location globally. However, many nations now have widespread concerns about reducing carbon emissions (Li et al., 2021: 1). It is important to reduce carbon emissions caused by temperature-controlled processing and storage facilities, especially in countries where electricity is generated from non-renewable sources. Additionally, it is important to promote the use of renewable energy sources, sometimes known as "cleaner energy," all over the world. According to research, air transport generates more greenhouse gas emissions per tonne of food than shipping does (Parashar et al., 2020: 3).

Everything suggests that we are in a time when implementing sustainability measures and communicating them is more important than ever. This is true both because doing so can improve the quality of the market's offerings and because the risks of not acting sustainably are highlighted by highlighting the negative externalities in the aforementioned environment (Mele et al., 2019: 2). Marketers now face both exciting potential and surprising obstacles as a result of the rapid expansion of environmentally friendly products across practically all relevant sectors. Budgets for green product advertising have nearly tripled since 2006 and have almost doubled over the past 20 years (Kim et al., 2021: 31). Some academics contend that green marketing can endure if customers consistently buy and use green items, however marketing has traditionally been seen as incompatible with sustainability, and a disconnect between consumer attitudes and real green buying behaviour has been noted (Sharma, 2021: 1218). Furthermore, the research has typically viewed these practises in an overly altruistic manner and has not taken into account the prospect of taking the potential related costs (Mele et al., 2019: 3). Despite several policy proposals, it is important to comprehend and support consumer attitudes towards green marketing. Businesses that want to market green products to a wide range of consumers must understand what influences green purchasing. Self-interest, generosity, self-accountability, and even green packaging can predict and affect green purchases (Sharma, 2021: 1218).

2 Literature Review

2.1 Overview of Carbon Emissions in the South African Consumer Market

The complex environment created by the carbon emissions from the South African consumer market is entwined with the nation's distinctive economic features. South Africa's economy, which is one of the biggest and fastest-growing emerging economies in both Africa and the world, stands apart due to a number of unique characteristics (Joshua et al.,

2020: 18014). With a GDP of US$368.3 billion and a GDP per person of US$6,374, South Africa has the second-largest economy in Africa, only behind Nigeria. It is ranked fourteenth globally in terms of greenhouse gas emissions. This distinction is closely related to the country's heavy reliance on coal, which culminated in the commitment to peak emissions between 2020 and 2025, followed by a stabilisation phase of roughly ten years before beginning a drop (World Bank, 2020). Notably, it has the position of being the main source of CO2 emissions in Africa, contributing around 45% of the continent's total emissions and holding the seventh place globally. The country's GHG emissions in 2015 totalled 460 million tonnes of CO2 equivalents (MtCO2e).

Emissions have increased gradually over the past five decades, peaking at 490 MtCO2e in 2008 after reaching 250 MtCO2e in 1970. Emissions then showed an uneven decline, matching the country's economic development (World Bank, 2020). As a member of important bargaining groups like BASIC, which is made up of Brazil, India, and China, the African Community, the G7 + China, and the Coalition of Rainforest Nations (CfRN), South Africa plays a crucial role in international emissions negotiations (Joshua et al., 2020: 18014). On September 25, 2015, South Africa made a significant move by submitting its climate agreement to the Paris Climate Summit, demonstrating its commitment to efforts to reduce global emissions (Adebayo & Odugbesan, 2020: 9378). Although South Africa faces issues similar to those experienced by other African nations, such as high rates of poverty, unemployment, and income inequality, its predominately urban population brings special problems (Iqani, 2019: 230).

The issues associated with mobile technology, industrial emissions, and excessive fossil fuel use for lighting, heating, and cooking are all part of South Africa's effort to reduce the negative consequences of CO2e. While certain provinces continued to largely use solid fuels for cooking and economic activities, increasing the danger of fires and burn injuries, electricity consumption increased from 50 to 78% between 2002 and 2013. Surprisingly, South Africa was one of the world's top CO2 emitters in 2013, emitting 0.71 kg CO2 per US$ GDP (PPP) in 2005 (Ahmad & Khattak, 2019: 532). Shifts in consumer behaviour, notably pronounced in developed countries, demonstrate a tendency towards buying low-carbon items, driven by an increased attention on both product quality and environmental features, in response to mounting concerns surrounding GHG emissions. The competitiveness of low-carbon products is predicted to increase as a result of this trend (Zhang et al., 2021: 19971). This effort to reduce emissions in line with customer preferences poses a difficult task because it requires juggling environmental concerns with immediate economic gain (Zhang et al., 2021: 19971).

Consumption has a significant impact on economic growth and development due to its fundamental position in the GDP. Growing consumption fuels both local and international industrial output, which in turn stimulates increased energy and capital goods use and, ultimately, higher CO2e emissions (Ahmad & Khattak, 2019: 531). Pressure to minimise carbon emissions is placed on supply chain businesses by the rising demand for low-carbon goods, which reflects changing customer tastes, and stricter government carbon emission rules. As a result, these organisations look for solutions to reduce emissions effectively, which may help them become more competitive (Zhang et al., 2021: 19971). The consumer market in South Africa has changed over the past 20 years, with the urban population growing, the average age dropping to 26.3 years, and spending rising. Notably, organisations must adopt customised strategies in order to successfully engage this demographic as the black middle class grows and acquires significant purchasing power (Berndt et al., 2021: 8). The majority of people live in highly populated cities and small towns, necessitating more complex analysis of the problems with air quality and carbon emissions than simply attributing them to houses and businesses (Iqani, 2019: 230).

2.2 *Climate Change Awareness and Consumer Behaviour*

Consumers' environmental consciousness is a critical market-driven factor that facilitates the environmental sustainability of the supply chain (Hong & Guo, 2019: 157). Consumers are increasingly considering "climate-friendly options" in response to the growing awareness of problems caused by externalities related to corporate activities and irresponsible consumption practises, such as global warming, air and water pollution, and resource depletion (Mele et al., 2019: 2). Consumer awareness and perceptions are crucial factors in the early stages of the purchasing decision-making process. A positive perception of a product does not guarantee its purchase. Various factors influence the buying decision (Ketelsen et al., 2020: 2). Understanding the effects of consumer behaviour on food purchases is important due to the significant influence consumer choice has on the environmental attributes of food in the market (Woo & Kim, 2019: 2). Consumers demonstrate their environmental responsibility by choosing to purchase products that are environmentally friendly. These products provide utility through their functional and environmental attributes.

Consumers are believed to vary in their evaluation of functional attributes but tend to have similar preferences for environmental attributes. Consumers exhibit varying evaluations of the functional attributes of a product, but they consistently prioritise environmental consciousness (Hong & Guo, 2019: 157). Consumers' environmental concerns and the resulting consequences of their actions are apparent. However, it is evident that existing consumption patterns do not consistently result in concrete actions and changes in consumption behaviours. Enhanced access to reliable information increases public environmental awareness and enables individuals to integrate environmental considerations into their daily choices. Understanding consumer decision-making and information utilisation is crucial for developing strategies to reduce the environmental impact of consumption patterns (OECD, 2020). Consumer behaviour refers to the active participation of individuals in the process of acquiring, consuming, and disposing of goods and services. Behaviour can be influenced by various factors, such as social groups (e.g. friends' influence on clothing choices) or organisations. Consumer behaviour's impact on society is considered significant because it affects the economy, society, and environment (Junior et al., 2019: 229).

Moreover, Behaviour change encompasses more than just individual consumer actions, such as buying low-carbon products, recycling, and reducing meat consumption. It also extends to the various roles and contexts individuals occupy, including their roles as community members, organisation participants, and citizens capable of influencing policies (Whitmarsh et al., 2021: 77). On the other hand, consumer attitude refers to the positive or negative feelings that influence the desire to acquire products or objects. A green attitude plays a moderating role in the relationship between Green Product Value (GPV) and green behaviours (Woo & Kim, 2019: 3). Some consumers are changing their buying behaviour to reduce the impact of their consumption habits over the environment, choosing an environment-friendly consumption behaviour, often called sustainable consumption. Sustainable consumers are those who give importance to quality and price and choose products and brands that adopt conservation practices and display environmental concern (Junior et al., 2019: 227). Therefore, green marketing is confirmed to be an effective tool to turn consumers' green awareness into actual purchasing behaviour (Hong & Guo, 2019: 155).

2.3 Marketing's Contribution to Carbon Reduction

Marketing encompasses the activities, institutions, and processes involved in the creation, communication, delivery, and exchange of valuable offerings for customers, clients, partners, and society. Marketing plays a significant role in influencing a company's performance. It is defined as an organisational orientation with seven dimensions: pro-activeness, opportunity focus, calculated risk-taking, innovativeness, customer intensity, resource leveraging, and value creation (Dushi et al., 2019: 86). Kotler (2020: 11) posits that the concept of marketing is a value creation discipline that creates value endeavours to fulfil human needs through the creation of value. The marketer determines the product features and services that will provide value. Marketers determine prices that generate value during the exchange process. Marketers select distribution channels that enhance accessibility and convenience for consumers. Marketers select messages that effectively communicate the value generated by their offerings. Most marketing definitions categorise marketing activities into four main areas: product, place, price, and promotion. These areas are commonly referred to as the marketing mix or 4Ps (Dushi et al., 2019: 88).

The retailer acquires the environmentally friendly product from the manufacturer and subsequently offers it to consumers within its market. The business utilises sustainable marketing strategies, such as environmental advertisements and consumer environmental education, to promote their environmentally friendly product. Businesses determine their sustainable marketing efforts by considering the trade-off between the costs and revenues associated with their marketing investments (Hong & Guo, 2019: 156). Li et al. (2021: 3) argue that businesses acquire products from manufacturers and sell them to environmentally conscious consumers (Fig. 1).

Product consumption can lead to the production of packaging waste, which is a significant adverse effect. In 2015, the global packaging market was estimated to be 4300 billion units, with 73% of these units being allocated for food and beverage purposes. One potential solution to address

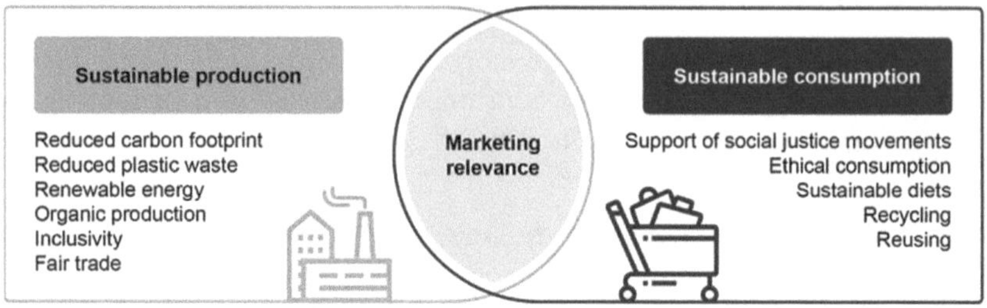

Fig. 1 Marketing's contribution to carbon emissions (*Source* Kabi, 2021: 40)

the issue of waste is the implementation of eco-friendly food packaging. Sustainable packaging refers to packaging that exhibits a relatively minimal environmental impact, as evaluated through life cycle assessment models. Sustainable packaging refers to the effort to minimise the environmental impact of a product by modifying its packaging, such as by utilising materials that are more environmentally friendly (Ketelsen et al., 2020: 2). The choice of packaging is typically driven by the goal of reducing costs, which has implications for both direct costs (such as purchasing and waste management) and indirect costs (including packing, handling, storing, and transporting). Effective packaging plays a crucial role in facilitating the storage, handling, transportation, and marketing of products (Yim et al., 2021: 32).

Green products are designed to meet the demands of environmentally conscious consumers. They are distinguished from conventional products by features such as unique packaging and environmental certifications that verify their eco-friendly nature. Certain products may appear to possess green characteristics, but in reality, they may lack such attributes. According to Junior et al. (2019: 227), these products are labelled as having an "environmental makeup" or are classified as greenwashing products. Sustainable packaging encompasses three levels: raw materials, production processes, and waste management. According to Ketelsen et al. (2020: 2), packaging should be biodegradable and capable of being reused at the end of its life cycle. In environmental claims, messages often incorporate quantifiable figures, such as "our battery, which is environmentally friendly, consists of 30% recycled materials." Concrete claims in advertising have been found to be more effective than abstract claims in enhancing credibility and memory retention. Consumers are generally more trusting of advertising claims that can be easily verified with objective and detailed information, rather than abstract claims (Yim et al., 2021: 32).

2.4 *Resistance to Climate-friendly Behaviour*

Globally, businesses fulfil their environmental obligations by offering and promoting eco-friendly products to consumers through sustainable marketing strategies to promote awareness and interest in environmentally friendly products (Hong & Guo, 2019: 157). However, there is a prevailing concern about the inherent disparities between developed and developing economies. Developing economies are currently starting to undergo the process of economic development that developed economies have already experienced. The main challenges faced by developing economies are poverty, deprivation, high unemployment, and limited technological advancements. One interpretation of sustainable production in this context is the improvement of local production capacity with the goal of generating employment opportunities and decreasing expenses (Roy & Singh, 2017: 226).

The key aspects to consider are the different challenges businesses face depending on the type of economy, the significance of a systematic approach that incorporates various disciplines and experts, the influence of consumer behaviour on business strategies, the importance of designing for sustainability and promoting system innovation, the role of government and policy frameworks, the impact of local communities on production and consumption preferences, and the role of technological advancements in facilitating collaborative responses (Roy & Singh, 2017: 226). Yet, this implementation of more comprehensive sustainable supply chain and marketing practices is associated with increased expenses including technology investment which, typically involves a higher initial investment compared to marketing efforts. This suggests that the cost rate for technology development often exceeds that of sustainable marketing (Hong & Guo, 2019: 157). On the other hand, various factors, such as health concerns, environmental conservation, the adoption of sustainable practises, conformity to societal expectations, and personal values, play a significant role in influencing consumers' decision to purchase eco-friendly products (Kumar et al., 2021: 2).

Furthermore, income and wealth disparities are other factors that influence decisions, as well as the ability to move geographically. The historical analysis demonstrates that economic development has resulted in both beneficiaries and disadvantaged individuals and nations. Income and wealth inequality lead to market segmentation opportunities and discrimination. Greater geographic mobility drives individuals to pursue better economic opportunities, resulting in immigration from economically disadvantaged countries to more prosperous ones. This phenomenon presents both challenges and opportunities related to diversity and inclusion (Trust, 2020: 17). Exorbitant costs within the organic market can function as a significant determinant for consumers to evaluate the dependability of a product. The significance of the value for money concept lies in its influence on consumers' positive attitudes towards green food products (Woo & Kim, 2019: 3).

Pricing policies influence incentives towards low-carbon alternatives or away from high-carbon options. Congestion charging, like other examples, has proven effective in reducing car usage and promoting the use of public transport and low-emission vehicles. Implementing such charges can disrupt automatic behaviours, leading to more deliberate and thoughtful purchase decisions (Whitmarsh et al., 2021: 78). This implies that product pricing is a crucial factor in the analysis of the consumer decision-making process. Previous studies have shown that the cost of environmentally friendly products affects the inclination of consumers to embrace such products (Kumar et al., 2021: 4). Trust (2020: 22) also suggests that wealth inequality and social instability have the potential to hinder economic growth and slow down the adoption of low-carbon. Different levels of wealth present unique opportunities within various economic strata from a marketing standpoint. Luxury markets are particularly important in situations marked by wealth inequality.

3 Methodology

This chapter follows a systematic review approach, based on the updated PRISMA 2020 (Preferred Reporting Items for Systematic Reviews and thematic synthesis) guidelines which were developed through a robust process involving literature reviews, expert consultation, and consensus building to improve transparency and completeness in systematic reviews (Page et al., 2021). The aim was to gather and examine published research on how marketing can help reduce carbon emissions in the South African consumer market. A wide search was carried out using several academic databases, including Scopus, Web of Science, ScienceDirect, SpringerLink, and Google Scholar. The search covered studies published between January 2017 and April 2024. Key terms were combined using Boolean operators, such as "climate change" AND "marketing", "carbon emissions" AND "South Africa", "green marketing", and "consumer behaviour".

Table 1 Above provides a structured overview of how this study adheres to the PRISMA 2020 guidelines for conducting systematic reviews. It highlights the key components covered such as inclusion and exclusion criteria, data sources, search strategy, selection process, and synthesis methods ensuring transparency and rigour. This table helps to validate the methodological soundness and reproducibility of the review process (Fig. 2)

A total of 542 records were initially identified through database searches. After removing duplicates, 412 records were screened by title and abstract. Of these, 280 were excluded for not meeting the inclusion criteria. The remaining 132 full-text articles were reviewed for eligibility, and 90 were excluded due to irrelevance or lack of focus on marketing and sustainability. Ultimately, 42 studies were included in the final synthesis, providing the evidence base for this review. Information from the selected

Table 1 Summary of PRISMA 2020 criteria addressed

PRISMA component	*How addressed in this review*
Title	Clearly identified as a systematic review
Abstract	Structured summary included key elements
Rationale	Explained within the South African carbon emissions context
Objectives	Stated to assess the role of marketing in reducing carbon emissions
Eligibility criteria	Inclusion of peer-reviewed articles from 2017 to 2024 focused on South Africa
Information sources	Peer-reviewed literature was retrieved from five databases: Scopus, Web of Science, ScienceDirect, SpringerLink, and Google Scholar. Search terms included combinations of "carbon emissions," "consumer behaviour," "marketing," "South Africa," and "climate change." Boolean operators and filters (e.g. publication date range 2017–2024) were applied to refine the results
Selection process	Titles, abstracts, and full-texts were screened independently by the author for relevance and eligibility
Data collection	Key themes, concepts, and findings were extracted using a predefined coding sheet
Synthesis	A narrative and thematic synthesis approach was used due to methodological heterogeneity across included studies

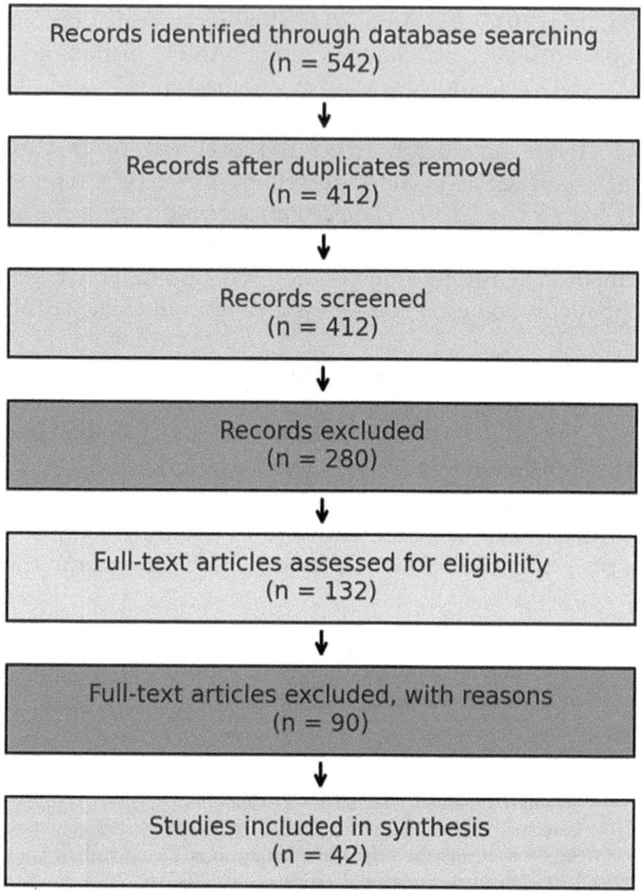

Fig. 2 PRISMA 2020 flowchart illustrating article screening and selection process (*Source* Authors' own creation)

studies was gathered using a coding sheet. This helped record details such as the author, year, aim of the study, method used, sample size, and key findings. A thematic analysis was then used to group the findings under common themes. This included consumer awareness, the effect of marketing on behaviour, barriers to sustainable choices, price and affordability issues, and the overall role of marketing in reducing carbon emissions. Because the studies were different in their methods and focus,

Table 2 Inclusion and exclusion criteria for systematic review

Inclusion criteria	*Exclusion criteria*
Peer-reviewed journal articles published between 2017 and 2024	Publications outside the date range of 2017 to 2024
Studies focusing on marketing, consumer behaviour, and climate change in the South African context	Studies unrelated to South Africa or not addressing marketing/consumer behaviour
Articles written in English	Non-English articles
Empirical or conceptual studies with relevant findings on carbon emissions, sustainability, and consumption patterns	Opinion pieces, editorials, or sources lacking methodological rigour
Accessible full-text articles	Abstract-only entries or inaccessible full-texts

a meta-analysis could not be done. Instead, the results were presented in a descriptive and narrative way (Table 2).

Studies were included if they were peer-reviewed, published between 2017 and 2024, written in English, and focused on marketing, consumer behaviour, or sustainability within the South African context. Articles were excluded if they lacked academic rigour, focused solely on technical or policy aspects of emissions without linking to marketing or consumers, or were not accessible through academic databases. As a result, it is important to note that this review has some limitations. Also, because of the differences in study design and focus, it was not always possible to compare results directly. Despite this, the review offers a clear summary of current research and provides a useful foundation for future studies on marketing and climate action in South Africa.

4 Findings and Discussion

4.1 Summary of Key Findings

These studies looked at how marketing, consumer behaviour, and carbon emissions are connected in the South African context. Some focused on the country's high levels of carbon emissions. For example, Adebayo and Odugbesan (2021) used advanced modelling to show how energy use affects emissions, while Ahmad and Khattak (2020) linked rising consumption to increased CO_2 levels. Marketing research added useful insights. Kotler (2020) and Trust (2020) discussed how marketing can help create value in ways that support sustainability. Dushi et al. (2019) explored how small businesses use marketing strategies, and Kabi (2021) highlighted how marketing in South Africa can promote sustainable behaviour when it is done with local needs in mind.

Studies on consumers showed how people think and act about the environment. Iqani (2019) explained how influencers shape what people buy, while Sharma (2021) reviewed green marketing strategies that aim to shift behaviour. Whitmarsh et al. (2021) added that real behaviour change needs more than awareness, it also needs supportive systems and policies. Other researchers looked at packaging and product supply chains. Hong and Guo (2019), Ketelsen et al. (2020), and Liu et al. (2021) showed that consumers care about the environment, but still face challenges such as price, access, and trust in green labels. Finally, broader economic studies from Joshua et al. (2020) and Hong et al. (2021) helped link consumer behaviour to national patterns like urban growth and investment trends.

Table 3 presents a summary of key studies reviewed in this chapter, grouped under major themes. It highlights how each theme reflects specific areas of academic inquiry related to marketing and carbon emission reduction within the South African consumer context. The results are organised thematically to highlight major areas of focus across the selected studies.

Figure 3 presents the frequency of each theme identified across the reviewed literature, offering a visual summary of the most researched topics. It shows that consumer awareness and the behavioural impact of marketing received the most attention, while affordability and low-carbon branding remain less explored.

4.2 *Discussion of Key Findings*

4.2.1 Consumer Awareness

The reviewed literature consistently highlights the importance of consumer awareness in driving environmentally responsible behaviour. Studies show that as individuals become more informed about the effects of climate change and the environmental costs of consumption, they are more likely to support eco-friendly products and practices (Hong & Guo, 2019; Junior et al., 2019). Environmental knowledge positively shapes attitudes, and many consumers express concern for sustainability and a willingness to act accordingly. However, a significant gap exists between consumers' expressed values and their actual behaviour a phenomenon

Table 3 Summary of systematic literature review

Theme	*Key studies*	*Main findings*
1. Overview of carbon emissions in South Africa	Joshua et al. (2020); World Bank (2020); Adebayo and Odugbesan (2020); Ahmad and Khattak (2019); Iqani (2019)	South Africa has high CO2 emissions due to coal dependency; urbanisation and inequality complicate sustainability transitions
2. Climate change awareness and consumer behaviour	Hong and Guo (2019); Mele et al. (2019); Ketelsen et al. (2020); OECD (2020); Woo and Kim (2019); Whitmarsh et al. (2021); Junior et al. (2019)	Consumers show increasing climate awareness, but gaps exist between attitudes and actual behaviour. Access to information and social context are important factors
3. Marketing's contribution to carbon reduction	Kotler (2020); Dushi et al. (2019); Li et al. (2021); Ketelsen et al. (2020); Yim et al. (2021); Junior et al. (2019)	Marketing strategies such as green packaging, eco-labelling, and environmental claims can influence behaviour; credibility and value delivery are crucial
4. Resistance to climate-friendly behaviour	Roy and Singh (2017); Kumar et al. (2021); Trust (2020); Woo and Kim (2019); Whitmarsh et al. (2021)	Affordability, wealth inequality, and skepticism about green products affect adoption; behaviour is shaped by context, incentives, and social dynamics

Source Authors' own creation

known as the behaviour intention gap. For example, while many South Africans report concern about climate change, fewer make consistent green purchases (OECD, 2020; Whitmarsh et al., 2021). This disconnect suggests that awareness alone is not enough. Other factors such as convenience, pricing, and trust in product claims often override good intentions. Therefore, while raising awareness remains essential, it must be supported by marketing strategies that enable and simplify green decision-making.

4.2.2 *Effect of Marketing on Behaviour*

Marketing plays a crucial role in translating awareness into action. Studies in the review reveal that eco-labelling, green branding, and environmental advertising can significantly influence consumer perceptions and choices

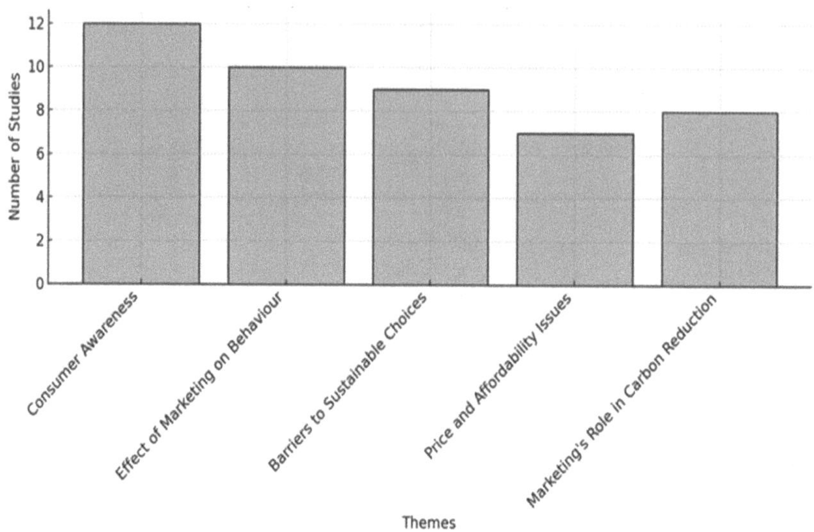

Fig. 3 Frequency of themes identified in systematic review

(Ketelsen et al., 2020; Yim et al., 2021). Informational claims such as recycled content percentages or biodegradable packaging are found to be more effective than vague slogans, as they enhance credibility and help consumers make informed decisions.

However, there is growing concern about the practice of greenwashing where companies exaggerate or fabricate their environmental credentials. Such tactics risk undermining trust in green marketing and may discourage environmentally conscious consumers. Clear standards, third-party certifications, and transparent communication are needed to ensure marketing genuinely supports sustainability goals (Hong & Guo, 2019; Junior et al., 2019).

4.2.3 Barriers to Sustainable Choices

Despite rising environmental awareness, many consumers in South Africa still face practical and psychological barriers to adopting sustainable practices. These include limited access to affordable green products, low confidence in environmental claims, and habitual consumption behaviours that resist change (Kumar et al., 2021; Roy & Singh, 2017). In lower-income communities, sustainability is often viewed as a luxury concern, with basic needs taking priority over ecological considerations. Marketers must address these challenges by designing inclusive strategies that do not assume high levels of disposable income or environmental literacy. Policy interventions, such as product subsidies or public education campaigns, may also help bridge these gaps. Encouragingly, when businesses and policymakers work together, the adoption of sustainable behaviour becomes more achievable across all segments of society.

4.2.4 *Price and Affordability*

Price remains one of the most influential factors shaping consumer decisions regarding green products. The review highlights that although many consumers express willingness to support sustainable options, high costs often deter follow-through (Trust, 2020; Woo & Kim, 2019). Green products are commonly priced at a premium, reflecting higher production costs and limited economies of scale. Wealth inequality in South Africa further deepens this divide. For many households, eco-friendly choices are simply unaffordable, making sustainability appear exclusive or elitist. Marketing strategies must therefore communicate value for money and consider pricing models that allow wider access. Moreover, policies that incentivise low-carbon consumption such as tax rebates or discounts for sustainable purchases can ease the financial burden and promote behavioural change.

4.2.5 *Marketing's Role in Carbon Reduction*

Finally, the literature confirms that marketing can be a strategic tool for promoting carbon reduction. Sustainable marketing does not only inform consumers. It shapes preferences, creates demand for eco-friendly innovation, and redefines value around long-term environmental impact (Kotler, 2020; Li et al., 2021). Companies are increasingly integrating sustainability into their brand identity, using packaging, product design, and messaging to signal low-carbon benefits. Green product innovation, supported by marketing, helps consumers align their values with their purchases. However, this requires more than surface-level messaging; authenticity and consistent sustainability efforts across the value chain are essential. When used ethically and creatively, marketing can become a powerful force in guiding the South African consumer market towards more climate-resilient behaviours.

5 Conclusions and Recommendations

This systematic review highlights the central role marketing can play in shaping environmentally responsible consumer behaviour and contributing to carbon emission reduction. While consumer awareness and concern about climate change are increasing, this has not consistently translated into widespread sustainable purchasing behaviour. The gap between environmental intentions and actions is often due to affordability barriers, scepticism about green claims, and limited access to clear and actionable information. The review further reveals that while businesses often benefit from government incentives and policy support for sustainability, similar measures for consumers are either lacking or poorly communicated. There is little evidence of structured efforts to empower or incentivise individual consumers to adopt low-carbon choices. This asymmetry limits the potential for large-scale behavioural shifts within the consumer base, particularly in a country marked by inequality and uneven access to sustainable products. In response to these findings, the following recommendations are proposed:

- Advocate for Green Marketing:
  Given the positive perception of eco-conscious brands, firms should continue investing in transparent and credible green marketing efforts. Strategies should prioritise trustworthy eco-labelling, sustainability-focused advertising, and authentic brand storytelling to counteract greenwashing and foster deeper engagement.
- Ensuring Affordability of Sustainable Products:
  Sustainable goods remain cost-prohibitive for many consumers. Companies should explore cost-reduction strategies such as improved production efficiencies and value-chain collaborations to decrease prices without compromising product quality. Simultaneously, policymakers should introduce direct incentives or subsidies for consumers such as vouchers, rebates, or reduced taxes on eco-friendly products to make sustainable choices more financially accessible.
- Promote Social Discourse on Climate Change:
  Despite rising awareness, climate-related conversations remain rare in daily social interactions. Marketing campaigns, educational content, and social media initiatives can help normalise discussions

around sustainability. Encouraging peer influence and social proof can strengthen pro-environmental norms and behaviours.

- Enhancing Public Knowledge of Government Initiatives:

  Although many companies benefit from sustainability incentives, consumer-targeted schemes are either non-existent or poorly communicated. Governments should broaden their climate action strategies to include awareness campaigns and financial incentives aimed directly at individuals. Clear, relatable messaging about available programmes and the environmental benefits of low-carbon living could drive behavioural change at scale.

References

Abbasi, K. R., Adedoyin, F. F., Abbas, S., & Hussain, S. (2021). The role of renewable energy and globalization in reducing environmental degradation: Evidence from a novel quantile-on-quantile regression approach. *Resources Policy, 70*, Article 101941.

Adebayo, T. S., & Odugbesan, J. A. (2020). Modeling CO2 emissions in South Africa: Empirical evidence from ARDL-based bounds and wavelet coherence approaches. *Environmental Science and Pollution Research, 27*, 9378–9392.

Ahmad, M., & Khattak, S. I. (2019). Economic growth and environmental quality: An empirical study on South Asian countries. *Environmental Science and Pollution Research, 26*(5), 531–541.

Berndt, A., Bick, G., Boshoff, C., Chigada, J., Klopper, H. B., Maduku, D. K., Mpinganjira, M., Ndoro, T., Rugimbana, R., & Strydom, J. (2021). *Marketing: South African edition*. Oxford University Press.

Dushi, N., Dana, L. P., & Ramadani, V. (2019). Entrepreneurial marketing: A review of literature. *International Journal of Entrepreneurship and Small Business, 36*(1–2), 86–101.

Goworek, H., & McGoldrick, P. (2015). *Retail marketing management: Principles and practice*. Pearson Education.

Hong, I. B., & Guo, X. (2019). Sustainable marketing strategy and consumer perception: Evidence from the green retail sector. *Sustainability, 11*(1), 157.

Iqani, M. (2019). Consumption, media and the global south: Aspiration contested. *Consumption Markets and Culture, 22*(2), 230–234.

Joshua, U., Bekun, F. V., & Sarkodie, S. A. (2020). New insight into the causal linkage between economic expansion, FDI, coal consumption, pollutant emissions and urbanization in South Africa. *Environmental Science and Pollution Research, 27*(15), 18013–18024.

Junior, L. A. L., Martinez, M. P., Correa, C. M., Moura-Leite, R. C., & Da Silva, D. (2019). Green consumer behavior: A bibliometric analysis of its evolution. *Journal of Cleaner Production, 231*, 229–236.

Kabi, M. (2021). Marketing's role in promoting low-carbon behaviour: A South African perspective. *Journal of Environmental Marketing, 8*(1), 35–45.

Ketelsen, M., Janssen, M., & Hamm, U. (2020). Consumers' response to environmentally-friendly food packaging: A systematic review. *Journal of Cleaner Production, 254*, Article 120123.

Kim, Y., Yim, J., Kim, S., & Reeves, M. (2021). Green advertising: Evolution and impact on sustainable consumer behaviour. *Journal of Strategic Marketing, 29*(1), 31–45.

Kotler, P. (2020). *Marketing 5.0: Technology for humanity*. Wiley.

Kumar, B., Prakash, A., & Kumar, N. (2021). Green marketing and consumer behaviour: The role of value-driven green beliefs. *Journal of Cleaner Production, 312*, Article 127740.

Li, Y., Pan, W., Yang, L., Ma, C., & Zhou, X. (2021). Decarbonisation and economic growth: A global analysis. *Sustainability, 13*(1), 1.

Liu, X., Wahab, S., Hussain, M., Sun, Y., & Kirikkaleli, D. (2021). China carbon neutrality target: Revisiting FDItrade-innovation nexus with carbon emissions. *Journal of Environmental Management, 294*, 113043.

Mele, C., Gomez, C., & Garay, L. (2019). Green tourism marketing: A meta-analysis of the environmental impact. *Sustainability, 11*(24), 6912.

OECD. (2020). *Consumer data and sustainable consumption: OECD report*. OECD Publishing.

Page, M. J., McKenzie, J. E., Bossuyt, P. M., Boutron, I., Hoffmann, T. C., Mulrow, C. D., Shamseer, L., Tetzlaff, J. M., & Moher, D. (2021). Updating guidance for reporting systematic reviews: Development of the PRISMA 2020 statement. *Journal of Clinical Epidemiology, 134*, 103–112.

Parashar, A., Sood, M., & Agrawal, S. (2020). Environmental impacts of food transportation systems. *Transportation Research Part d: Transport and Environment, 86*, Article 102432.

Roy, V., & Singh, S. (2017). Prioritizing sustainable production indicators for the food-processing industry: An Indian case study. *Sustainable Production and Consumption, 12*, 226–242.

Sharma, E. (2021). Consumer responses to green advertising: A review and future directions. *International Journal of Consumer Studies, 45*(6), 1217–1220.

Strydom, J. (2016). *Introduction to marketing* (5th ed.). Juta.

Trust, E. (2020). *World inequality report: Bridging the divide*. Global Inequality Lab.

Whitmarsh, L., Poortinga, W., & Capstick, S. (2021). Behavioural insights and climate action. *Nature Climate Change, 11*, 77–84.

Woo, E., & Kim, Y. G. (2019). Consumer attitudes toward green products: A meta-analysis. *Sustainability, 11*(6), 1621.

World Bank. (2020). *South Africa's carbon intensity profile.* World Bank Reports.

Yim, M. Y. C., Kim, Y. K., & Lee, J. (2021). How to easily facilitate consumers' mental simulation through advertising: The effectiveness of self-referencing image dynamics on purchase intention. *International Journal of Advertising, 40*(5), 810–834.

Zhang, D., Liu, Z., Li, W., & Anwar, S. (2021). The dynamic impact of low-carbon products on CO2 emissions. *Environmental Science and Pollution Research, 28*, 19971–19985.

CHAPTER 12

Circular Economic Practices in BRICS: Challenges, Opportunities, and Policy Implications—A Systematic Review

Aifani Confidence Tahulela, Abraham Adeniran, and Fulufhelo Netswera

1 Introduction

The concept of a circular economy (CE) has gained significant attention in recent years as a sustainable approach to economic development (Ottoni et al., 2020). It offers an alternative to the linear "take-make-dispose" model by emphasizing the reduction, reuse, and recycling of materials, as well as the regeneration of natural systems (Kurniawan et al., 2022). A CE aims to decouple economic growth from resource consumption by promoting the efficient use of resources and maximizing their value through reuse, recycling, and regeneration. As nations worldwide

A. C. Tahulela (✉) · A. Adeniran
Department of Public Management, Durban University of Technology, Durban, South Africa
e-mail: aifaniT@dut.ac.za

F. Netswera
Durban University of Technology, Durban, South Africa

N. Tshikovhi et al. (eds.), *The BRICS Imperative*,
https://doi.org/10.1007/978-3-032-05932-1_12

strive to achieve the United Nations' Sustainable Development Goals (SDGs) and address environmental challenges, understanding the implementation of CE practices becomes crucial (Kurniawan et al., 2022. In the context of the BRICS nations (Brazil, Russia, India, China, and South Africa), the adoption of circular economic practices presents unique challenges, opportunities, and policy implications. The BRICS countries, characterized by their large populations, rapid economic growth, and diverse industrial sectors, play a crucial role in the global economy and are increasingly focusing on sustainable development (Sohal et al., 2022). According to Fedotkina et al. (2019); Dorokhina et al. (2021); Kostygova (2019); Vasilenok et al. (2021) the BRICS nations face common challenges in managing their resource consumption and waste generation. As shown in the graph below (Fig. 1), the percentage of waste recycled in BRICS nations is still relatively low; a challenge these nations should address to transition into circular economies. Rapid urbanization and industrialization have resulted in increased demand for resources and an upsurge in waste streams. Additionally, the reliance on traditional linear economic models, which prioritize extraction, production, consumption, and disposal, has further exacerbated these challenges.

The percentage waste recycled in BRICS in 2020 shows China to have the highest recycling rate (38%) followed by Brazil (29%), India (24%)

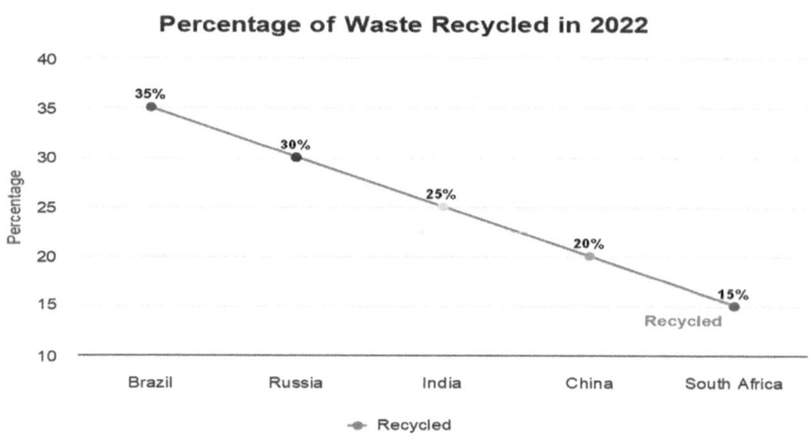

Fig. 1 Percentage of waste recycled in 2022 (*Source* World Bank (World Development Indicators [2023])

while Russia and South Africa have the lowest recycling rates, i.e., 14% and 12% respectively According to Maiurova et al. (2022), the BRICS nations possess significant opportunities for circular economic implementation because their large and growing consumer markets provide a strong impetus for adopting sustainable production and consumption practices. Furthermore, these countries are rich in natural resources, which can be harnessed through circular approaches to ensure their long-term availability and reduce dependence on imports (Guimaraes Araújo et al., 2021). CE policies should incentivize resource efficiency, support innovation and technology transfer, encourage collaboration between public and private sectors, and foster consumer awareness and behavioral change (Devendran et al., 2023; Guimaraes Araújo et al., 2021; Jesus et al., 2021).

This systematic review on BRICS nations recognizes these countries' growing political influence on the global stage and their significant contribution to global economic growth. However, their rapid industrialization and consumption patterns have led to environmental degradation and resource depletion (Kandachar & Halme, 2019). Embracing circular economic practices presents both challenges and opportunities for these nations if they should transit into more sustainable development pathways (Kalchenko et al., 2019; Maiurova et al., 2022). The rationale for conducting this systematic review lies in the need to assess the current state of circular economic practices, identify associated challenges and opportunities, and derive policy implications for BRICS nations.

While individual case studies and research papers have examined specific aspects of CE adoption within BRICS nations, a comprehensive review synthesizing existing knowledge should yield a more holistic understanding of this subject matter. Such a systematic approach facilitates the identification of literature gaps, best practices, and evidence-based policy recommendations. Several studies have explored diverse facets of CE practices within BRICS nations, providing valuable insights into associated challenges, opportunities, and policy implications. For example, studies conducted by Zeng and Li (2020); Kurniawan et al. (2022) on circular economic practices in China, focused on industrial symbiosis and waste recycling. The findings highlighted notable progress in promoting CE initiatives in China but also identified challenges such as fragmented policies, inadequate stakeholder coordination, and insufficient monitoring and enforcement mechanisms. However, this study primarily concentrates on China, lacking a comprehensive analysis of CE practices across all BRICS nations.

The research question guiding this systematic review is: *What are the challenges and opportunities associated with the implementation of CE practices within BRICS nations, and what are the policy implications for fostering sustainable development in these countries?* By addressing this research question, we aim to provide insights into the current state of CE practices within BRICS nations, identify barriers to implementation, and suggest potential policy interventions to expedite the transition toward a CE.

2 Methods

This study adopted the Systematic Reviews and Meta-Analyses (PRISMA) methodology ensuring a comprehensive and unbiased evaluation of previous studies. The PRISMA approach was particularly effective in ensuring the relevance and accuracy of the selected articles (Page et al., 2021). The PRISMA methodology was notably proficient in ascertaining the relevance and precision of the chosen articles. The process commenced with a meticulous search of databases to gather relevant papers. Utilizing the Scopus database, an advanced search was conducted with Boolean operators in the title, abstract, and keywords fields. The search combined the core concept of the circular economy with BRICS and country-specific terms, as well as implementation and policy-related themes covering practices, initiatives, challenges, opportunities, drivers, and governance. The exact search string used was: TITLE-ABS-KEY (("circular econom*" OR "circularity" OR "closed-loop" OR "resource efficien*" OR "waste minim*" OR "urban mining") AND (BRICS OR "BRICS nations" OR Brazil OR Russia OR India OR China OR "South Africa") AND (practic* OR implement* OR initiative* OR strateg* OR framework* OR transition*) AND (challeng* OR barrier* OR constraint* OR opportunit* OR driver* OR enabling) AND (policy OR polic* OR regulation* OR governance OR institution* OR "policy implication*")). This strategic use of search terms ensured a broad yet targeted exploration of the relevant literature (Fig. 2).

Our initial search through the database yielded 1,210 articles, which, after careful screening, was reduced to 494. Consequently, we selected 50 articles, the criteria where that they were written in in English, had undergone peer-review, and were recent enough to truly mirror the prevailing

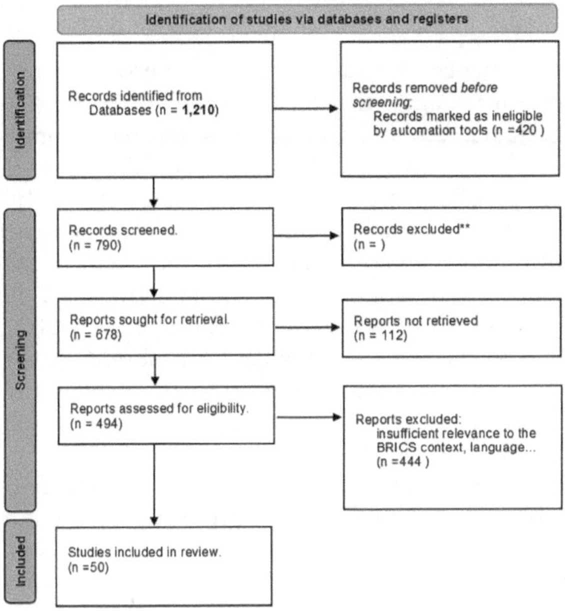

Fig. 2 PRISMA flow (*Source* generated by authors)

circumstances. Articles not fitting the criterion where discarded. Sampled articles were subjected to an extensive screening process where the title and abstract were initially evaluated for relevance, followed by a thorough perusal of the full text. Full details of this studies can be found in appendix 1 to 5.

3 Discussion of Results

Findings from the reviewed literature are categorized as follows; CE Practices in the BRICS Nation; Challenges in Implementing CE in the BRICS; Opportunities for CE in the BRICS Nation; Technology and Innovation in CE in BRICS; and Policy Implications for CE.

3.1 CE Practices in the BRICS Nation

In terms of CE Practices, Literatures on circular economic practices across the BRICS nations suggest that in Brazil, there is a progressive inclination toward circular economic practices is apparent. Significant strides are evident in the bioenergy and bioeconomy sectors, with a focus on sustainable usage of residues (Coelho et al., 2021). Public organizations, including universities, have implemented Waste Management Plans (WMPs) contributing to eco-friendly waste management (Ottoni et al., 2020). Moreover, in retail, supermarkets employ Big Data predictive analytics to reduce food wastage (de Souza et al., 2021). Furthermore, strategic partnerships for biodigesters, lubricating oil re-refining, and waste management optimization are being explored, highlighting a concerted movement toward sustainable practices (Devendran et al., 2023; Guimaraes Araújo et al., 2021; Jesus et al., 2021).

Studies from Russia reveals that the country approach to the CE revolves largely around the recycling of production waste into secondary resources, and the adoption of sustainable business models as highlighted in studies by Fedotkina et al. (2019); Dorokhina et al. (2021); Kostygova (2019); Vasilenok et al. (2021). A notable strategy is the establishment of Eco-Industrial Parks (EIPs), which function on principles of industrial symbiosis (Abu-Qdais et al., 2022; Saha et al., 2021). The energy sector is embracing green technologies and circular economic principles, with digital transformation also being integrated to enhance waste collection and recycling efficiency (Kalchenko et al., 2019; Maiurova et al., 2022).

Studies from India also showcases a wide range of circular economic practices. While recycling is a significant aspect, the emphasis is placed on the broader "reduce, reuse, recycle" paradigm. The application of 'urban mining' in e-waste management is a unique example of this approach (Debnath et al., 2021). Moreover, successful CE business models, such as that employed by the Paperman Foundation, and the incorporation of smart practices within Smart Cities further underline the diversity of approaches (Hossain et al., 2022; Kapoor et al., 2022; Siddiqui & Pandit, 2021). Shifts in recycling concepts in the ELV sector and the implementation of CE in SMEs showcase India's commitment to sustainable practices (Molla et al., 2022; Sohal et al., 2022).

In China CE practices significant diversified, with a shared focus on waste reduction and resource optimization (Li et al., 2023; Tang et al., 2020; Zeng & Li, 2020). Efforts span from recycling to total redesign of production and consumption processes. Cities piloting CE programs illustrate these wide-ranging practices (Chen et al., 2020). Mandatory waste sorting and digital transformation of waste management are significant developments underlining China's aggressive CE push (Kurniawan et al., 2022; Steuer, 2021; Zhang et al., 2023).

South Africa's CE practices incorporate a wide array of methods and technologies. These include eco-innovation, green supply chain management, big data-driven supply chains, and AI-powered sustainable manufacturing (Bag et al., 2021, 2022). Concurrently, traditional practices of reuse, recycling, water conservation, and material recycling also play significant roles (Adeleke et al., 2021; Godfrey et al., 2021; Mativenga et al., 2017; Olatayo et al., 2021; van Zyl et al., 2022). Further, educational institutions foster sustainable practices such as recycling (Owojori et al., 2022), and remanufacturing has emerged as a strategy to prolong product lifecycles (Ohiomah et al., 2022). In conclusion, each of the BRICS nations displays a unique set of CE practices, adapting to their local contexts, while sharing a common goal toward sustainability.

3.2 Challenges in Implementing CE in the BRICS

In implementing CE practices across BRICS, diverse set of challenges that are tied to each country's unique context were identified from the literatures.

Brazil has made significant progress in embracing CE practices; however, challenges persist. From the reviewed studies Issues around optimizing existing practices, addressing urban inequalities, and comprehending the role of emerging digital technologies in sustainability where identified (Coelho et al., 2021; de Souza et al., 2021; Sehnem et al., 2020). Also, according to van der Velden et al. (2022), there is a noticeable lack of awareness among farming communities regarding the benefits of organic waste recycling. Moreover, the imprecision of laws about the correct disposal and treatment of biodegradable waste, and the lack of data on Brazilian policies that emphasize the CE, hamper progress (Coelho et al., 2021).

Russia, also faces significant obstacles in its path to a CE. The literatures under review reveals lack of a systematic approach to waste processing, an inadequate institutional base, and difficulties in defining roles within waste management ecosystems all pose challenges (Maryev et al., 2021). Other hurdles identified include limited stakeholder support for infrastructure development, a short-term profit mindset, significant waste generation in the industrial sector, and the need for economic assessments of waste utilization practices (Dorokhina et al., 2021; Fedotkina et al., 2019; Gadzhiev et al., 2020; Kalchenko et al., 2019; Kostygova, 2019; Vasilenok et al., 2021).

In India, transitioning from a 'take-make-dispose' model to a 'take-make-reduce' model is a significant issue (Kapoor et al., 2022). Challenges also abound in the realm of plastic waste management, with few potential solutions to the enormous problem (Hossain et al., 2022). The informal handling of e-waste (Debnath et al., 2021), limited governmental support for SMEs (Sohal et al., 2022), and pollution from untreated waste (Ghosh et al., 2021) serve as further barriers to the effective implementation of a CE.

In China, challenges are related to the excessive generation of municipal solid waste, environmental and public health concerns, supply chain disruptions, and regional resource shortages (Kurniawan et al., 2022; Tang et al., 2020; Zeng & Li, 2020). The implementation of waste sorting programs has encountered obstacles due to hasty planning and insufficient grassroots policy support (Zhang et al., 2023). Moreover, an initial economic slowdown in cities adopting CE approaches and the difficulty in adequately disclosing CE information to stakeholders add to the complexity (Chen et al., 2020; Kuo & Chang, 2021).

In South Africa, understanding the impacts of various practices, adopting AI in manufacturing, and managing waste effectively are key challenges (Adeleke et al., 2021; Bag et al., 2021, 2022; Botha et al., 2022). Issues with data accuracy and reliability, coupled with gaps in students' knowledge, can hinder progress (Owojori et al., 2022). Additionally, the lack of comprehensive data, assumptions in waste evaluation, and barriers to remanufacturing, such as quality concerns, policy limitations, and poor customer perception, further compound these challenges (Olatayo et al., 2021; Ohiomah et al., 2022).

In summary, while the BRICS nations are actively embracing the CE, each faces distinct challenges tied to their unique economic, social, and political contexts. Addressing these obstacles necessitates tailored strategies that account for local conditions and stakeholders' perspectives.

3.3 Opportunities for CE in the BRICS Nation

In the face of their unique challenges, the BRICS nations also possess a wealth of opportunities to leverage CE practices to their benefit.

In Brazil, there is significant potential for aligning CE practices with business model maturity (Sehnem et al., 2020), improving basic sanitation through sustainable use of residues (Coelho et al., 2021), and enhancing waste management in universities (Ottoni et al., 2022). Strategic location decisions and route optimization can further optimize waste management (Devendran et al., 2023; Jesus et al., 2021). Additionally, opportunities abound in loop closing in lubricating oil re-refining (Guimaraes Araújo et al., 2021) and diversifying income sources through closed-loop organic waste management systems (van der Velden et al., 2022).

In Russia, the expansion of new markets, the development of innovative solutions, and potential cost savings offer promising avenues for the growth of the CE (Fedotkina et al., 2019). Optimizing resource cycles could help establish a closed-loop economy, and the establishment of EIPs could enhance waste management (Dorokhina et al., 2021; Maryev et al., 2021). The industrial sector offers immense development opportunities through the rational use of resources (Kostygova, 2019). Digital technologies and renewable energy sources could further boost the country's CE (Gadzhiev et al., 2020; Maiurova et al., 2022; Vasilenok et al., 2021).

In India, the adoption of CE principles presents numerous opportunities that could also inform CE development in other countries (Kapoor et al., 2022; Hossain et al., 2022). Multi-stakeholder collaborations (Debnath et al., 2021), integration of CE principles within Smart City objectives (Siddiqui & Pandit, 2021), and the potential for waste management to become a profitable business model (Ghosh et al., 2021) all offer exciting prospects.

China also boasts numerous opportunities despite the challenges it faces. Improvements in resource output, consumption, and waste management offer significant potential (Tang et al., 2020; Zeng & Li, 2020), while cost reduction and technology innovation provide further benefits (Kurniawan et al., 2022). Reimagining production processes can have a positive impact on firms' financial performance (Li et al., 2023).

In South Africa, the transition toward a CE can offer benefits such as the integration of eco-innovation and green supply chain management to improve SME capabilities (Bag et al., 2022). The adoption of big data analytics and AI can enhance sustainable manufacturing and waste management (Adeleke et al., 2021; Bag et al., 2021). Additional opportunities lie in environmental education and participatory programs (Owojori et al., 2022), water conservation strategies (van Zyl et al., 2022), and cost reduction through recycling (Mativenga et al., 2017).

In summary, the opportunities present in the BRICS nations demonstrate the potential for positive transformation through the adoption of CE practices. The exploitation of these opportunities will however require thoughtful and inclusive strategies that recognize each country's unique context.

3.4 Technology and Innovation in CE in BRICS

Technology and innovation are pivotal elements in driving the transition to a CE in the BRICS nations, spanning various facets such as waste management, digital transformation, process optimization, and innovative business models.

In Brazil, digital technologies like Big Data predictive analytics, sensors, and augmented reality have been spotlighted for their potential to advance the CE, notably in reducing food waste (De Souza et al., 2021). Moreover, a combination of hierarchical algorithms with multicriteria analysis and Geographic Information Systems (GIS) and technologies such as Anaerobic Digestion, pyrolysis, hydroponics, and vermifiltration also offer substantial contribution to this transition (Jesus et al., 2021; van der Velden et al., 2022).

In Russia, innovations in infrastructure and process development for Eco-Industrial Parks (EIPs), adoption of environmentally friendly technologies, and the use of best available technologies are deemed essential (Dorokhina et al., 2021; Kalchenko et al., 2019; Maryev et al., 2021; Saha et al., 2021; Vasilenok et al., 2021). Furthermore, the digital transformation has been spotlighted for its potential to improve efficiency, including the use of smart technologies and convolutional neural network-based identification systems for waste collection, delivery, and processing (Maiurova et al., 2022).

In India, technological innovation is central to CE, reflected in innovative business models (Kapoor et al., 2022), the use of various waste utilization technologies (Ghosh et al., 2021), and the application of the TOPSIS-SIMOS approach in waste management (Kamble et al., 2021).

For China, technology and innovation underpin the CE transition, with the digitization of waste management playing a key role in waste reduction, reuse, and recycling (Kurniawan et al., 2022; Zhang et al., 2021). In particular, advances in Industry 4.0 technologies have facilitated smart waste management in supply chain operations.

While there are paucity of literature on CE in South Africa, however, studies such as Bag et al. (2022), revealed utilization of big data in supply chains the adoption of big data analytics-powered AI, furthermore, Botha et al. (2022) proposed decision support system for decentralized general waste management. In essence, the integration of technology and innovation in the CE in BRICS nations is a testament to the pivotal role of technological progress in facilitating sustainable and efficient resource management and optimization.

3.5 Policy Implication

The adoption of circular economic practices within BRICS nations holds great potential for addressing sustainability challenges, promoting resource efficiency, and fostering economic growth. To effectively harness these opportunities, there is need for policymakers to consider several key policy implications derived from a systematic review of circular economic practices. Here are the policy implications that should be considered:

1. Comprehensive CE Frameworks:
    - BRICS countries should prioritize the development and implementation of comprehensive Circular Economy (CE) policies that encompass various facets of sustainability, including waste management, recycling, and sustainable business practices.
    - Clear and detailed guidelines should be established for different industries and sectors, facilitating a smoother transition toward circular practices. This includes setting specific targets and milestones for each sector to achieve.
2. Mandatory Waste Management Plans:
    - Enforce the mandatory use of Waste Management Plans across industries and institutions to promote responsible resource utilization and waste reduction. These plans should be tailored to the specific needs and characteristics of each sector.

- Integrate waste reduction targets into these plans, ensuring that industries are held accountable for measurable progress toward reducing their environmental footprint.

3. Extended Producer Responsibility (EPR):

    - Implement the Extended Producer Responsibility (EPR) principle to hold producers accountable for the entire lifecycle of their products, from creation to disposal. This encourages manufacturers to consider the environmental impact of their products throughout their lifespan.
    - Encourage producers to design products with circularity in mind, emphasizing recyclability, reusability, and minimizing waste generation. Incentives such as reduced taxes or access to markets can be provided to manufacturers who excel in this regard.

4. Incentives for Green Technologies:

    - BRICS nations should introduce policies that provide strong incentives for the adoption of green technologies, energy-efficient practices, and environmentally friendly production methods. Financial incentives, tax breaks, and regulatory support can motivate industries to transition toward more sustainable operations.
    - Policymakers should also allocate funding for research and development in sustainable technologies, fostering innovation and promoting the creation of new green solutions within the BRICS nations.

5. Data-Driven Decision Making:

    - Promote the use of data-driven approaches in waste management and resource allocation by developing policies that encourage the collection and analysis of relevant data. Utilizing advanced analytics and technologies can lead to more informed and effective CE strategies.
    - Collaborate with industries and research institutions to gather data on resource flows, waste generation, and environmental impact. Sharing this information transparently can help identify opportunities for improvement and inform policy decisions.

4 Conclusion

This review suggests that the transition toward a CE within BRICS nations is complex and multifaceted. It offers valuable insights into the current landscape of sustainability efforts, although there are disparities in the depth and breadth of available literature for each country. It is evident that this literature is skewed toward China and Brazil, with a more limited representation from India, and an even scarcer presence from South Africa. This imbalance in the literature suggests that there is room for more comprehensive research and documentation of CE practices in all BRICS countries to provide a more holistic understanding. In terms of policy, the review identifies that China, Brazil, and India have developed relevant policies and practices related to the circular economy. China, in particular, has made significant strides in implementing CE policies. On the other hand, Russia and South Africa appear to have limited policies specifically focused on CE, indicating a potential area for policy development in these countries. Furthermore, the review identifies several indicators that can be used to assess the transition to a circular economy, they include; Resource Efficiency: Evaluating how efficiently resources are utilized and whether there is a reduction in resource consumption. Waste Reduction: Assessing the reduction in waste generation and increased recycling rates; Technological Innovation: Evaluating the adoption of innovative technologies that promote circularity; Extended Producer Responsibility (EPR): Analyzing whether EPR principles are in place to hold producers accountable for the lifecycle of their products; Data-Driven Decision Making: Assessing the use of data and analytics for informed decision-making in CE practices.

In conclusion, the systematic review of circular economic practices in the BRICS nations has provided valuable insights into the current state of CE, the challenges faced, the opportunities available, and the role of technology and innovation in driving the transition. The policy implications drawn from this review emphasize the importance of comprehensive CE frameworks, mandatory waste management plans, extended producer responsibility (EPR), incentives for green technologies, and data-driven decision-making. These policy recommendations aim to create an enabling environment for circular economic practices to flourish in the BRICS nations. It is crucial for policymakers in these countries to recognize the potential of CE in addressing sustainability challenges, promoting resource efficiency, and driving economic growth. By implementing these policies and working collaboratively with industries, academia, and other stakeholders, the BRICS nations can make significant strides toward achieving their sustainability goals and contributing to the global transition to a circular economy.

References

Abu-Qdais, H. A., & Kurbatova, A. I. (2022). The role of eco-industrial parks in promoting circular economy in Russia: A life cycle approach. *Sustainability (Switzerland), 14*(7).

Adeleke, O., Akinlabi, S. A., Jen, T. C., & Dunmade, I. (2021). Sustainable utilization of energy from waste: A review of potentials and challenges of Waste-to-energy in South Africa. *International Journal of Green Energy, 18*(14), 1550–1564.

Ahmed, J. U., Islam, Q. T., Ahmed, A., & Amin, S. B. (2023). Extending resource value-based circular economy business model in emerging economies: Lessons from India. *Business Perspectives and Research, 11*(2), 309–321.

Bag, S., Dhamija, P., Bryde, D. J., & Singh, R. K. (2022). Effect of eco-innovation on green supply chain management, circular economy capability, and performance of small and medium enterprises. *Journal of Business Research, 141*, 60–72.

Bag, S., Dhamija, P., Gupta, S., & Sivarajah, U. (2021). Examining the role of procurement 4.0 towards remanufacturing operations and circular economy. *Production Planning and Control*, *32*(16), 1368–1383.

Botha, M., Terblanche, S. E., & Luies, R. (2022). A decision support system for business development around decentralised waste utilisation in South Africa. *Cleaner Environmental Systems*, *7*.

Chen, F., & Yao, W. (2020). Evaluation of circular economy efficiency in eastern China. In Proceedings of *IOP Conference Series: Earth and Environmental Science*. https://doi.org/10.1088/1755-1315/453/1/012030

Coelho, S. T., Diaz-Chavez, R., Cortez, C. L., Perecin, D., Possetti, G. R. C., Rietow, J. C., & Silva, C. (2021). Circular economy in Brazil. In *Circular economy: Recent trends in global perspective* (pp. 459–496). https://doi.org/10.1007/978-981-16-0913-8_15

de Souza, F. F., Ferreira, M. B., Saraceni, A. V., Betim, L. M., Pereira, T. L., Petter, R. R. H., Pagani, R. N., de Resende, L. M. M., Pontes, J., & Piekarski, C. M. (2020). Temporal comparative analysis of industrial symbiosis in a business network: Opportunities of circular economy. *Sustainability (Switzerland), 12*(5), 1–18.

de Souza, M., Pereira, G. M., Lopes de Sousa Jabbour, A. B., Chiappetta Jabbour, C. J., Trento, L. R., Borchardt, M., & Zvirtes, L. (2021). A digitally enabled circular economy for mitigating food waste: Understanding innovative marketing strategies in the context of an emerging economy. *Technological Forecasting and Social Change, 173.*

Debnath, B., Das, A., & Das, A. (2021). Towards circular economy in e-waste management in India: Issues, challenges, and solutions. In *Circular economy and sustainability: Volume 2: Environmental Engineering* (pp. 523–543). https://doi.org/10.1016/B978-0-12-821664-4.00003-0

Devendran, A. A., Mainali, B., Khatiwada, D., Golzar, F., Mahapatra, K., & Toigo, C. H. (2023). Optimization of municipal waste streams in achieving urban circularity in the city of Curitiba, Brazil. *Sustainability (Switzerland), 15*(4).

Dorokhina, E. Y., & Kharchenko, S. G. (2021). Circular economy in Russia. In *Circular economy: Recent trends in global perspective* (pp. 309–327). https://doi.org/10.1007/978-981-16-0913-8_10

Fedotkina, O., Gorbashko, E., & Vatolkina, N. (2019). Circular economy in Russia: Drivers and barriers for waste management development. *Sustainability (Switzerland), 11*(20).

Gadzhiev, N. G., Murzak, N. A., Mitenkova, A. E., Skripkina, O. V., & Konovalenko, S. A. (2020). Problems of development of the circular economy as a factor in Russia's sustainable development. *South of Russia: Ecology, Development, 15*(3), 155–164.

Ghosh, S. K., Ghosh, S. K., & Baidya, R. (2021). Circular economy in India: Reduce, reuse, and recycle through policy framework. In *Circular economy: Recent trends in global perspective* (pp. 183–217). https://doi.org/10.1007/978-981-16-0913-8_6

Godfrey, L., Roman, H., Smout, S., Maserumule, R., Mpofu, A., Ryan, G., & Mokoena, K. (2021). Unlocking the opportunities of a circular economy in South Africa. In *Circular economy: Recent trends in global perspective*, pp. 145–180. https://doi.org/10.1007/978-981-16-0913-8_5

Guimaraes Araújo, M., Lovón-Canchumani, G., & Bechara Elabras Veiga, L. (2021). Circular economy for lubricating oils in Brazil. In *Proceedings of Springer proceedings in mathematics and statistics* (pp. 103–113). https://link.springer.com/chapter/10.1007/978-3-030-78570-3_8

Guzzo, D., Rodrigues, V. P., Pigosso, D. C. A., & Mascarenhas, J. (2022). Analysis of national policies for Circular Economy transitions: Modelling and simulating the Brazilian industrial agreement for electrical and electronic equipment. *Waste Management, 138*, 59–74.

Hossain, R., Islam, M. T., Shanker, R., Khan, D., Locock, K. E. S., Ghose, A., Schandl, H., Dhodapkar, R., & Sahajwalla, V. (2022). Plastic waste management in India: Challenges, opportunities, and roadmap for circular economy. *Sustainability (Switzerland), 14*(8).

Jesus, R. H. G. D., Barros, M. V., Salvador, R., Souza, J. T. D., Piekarski, C. M., & Francisco, A. C. D. (2021). Forming clusters based on strategic partnerships and circular economy for biogas production: A GIS analysis for optimal location. *Biomass and Bioenergy*, *150*.

Kalchenko, O., Evseeva, S., Evseeva, O., & Plis, K. (2019). Circular economy for the energy transition in Saint Petersburg, Russia. In*Proceedings of E3S Web of Conferences*. https://doi.org/10.1051/e3sconf/201911002030

Kamble, C. B., Raju, R., Vishnu, R., Rajkanth, R., & Pariatamby, A. (2021). A circular economy model for waste management in India. *Waste Management and Research, 39*(11), 1427–1436.

Kapoor, N., Jauhari, S., & Maheshwari, D. (2022). Practicing circular economy in India: Insights of business models on the three paradigms-reduce, reuse, and recycle. In *Handbook of research on developing circular, digital, and green economies in Asia* (pp. 179–196). https://doi.org/10.4018/978-1-7998-8678-5.ch010

Kostygova, L. (2019). Use of waste of production—The perspective direction of development of circular economy in the industry of Russia. In *Proceedings of International Multidisciplinary Scientific GeoConference Surveying Geology and Mining Ecology Management, SGEM* (pp. 911–918). https://www.epslibrary.at/sgem_jresearch_publication_view.php

Kuo, L., & Chang, B. G. (2021). The affecting factors of circular economy information and its impact on corporate economic sustainability-Evidence from China. *Sustainable Production and Consumption, 27*, 986–997.

Kurniawan, T. A., Maiurova, A., Kustikova, M., Bykovskaia, E., Othman, M. H. D., & Goh, H. H. (2022). Accelerating sustainability transition in St. Petersburg (Russia) through digitalization-based circular economy in waste recycling industry: A strategy to promote carbon neutrality in era of Industry 4.0. *Journal of Cleaner Production, 363*. https://doi.org/10.1016/j.jclepro.2022.132452

Li, Y., Hu, Y., Li, L., Zheng, J., Yin, Y., & Fu, S. (2023). Drivers and outcomes of circular economy implementation: Evidence from China. *Industrial Management & Data Systems, 123*(4), 1178–1197. https://doi.org/10.1108/imds-05-2022-0267

Li, Y., Zambrano, F., Wang, Y., & Marquez, R. (2022). How China's foreign waste ban will reshape the U.S. recycling supply chain: Economic and environmental considerations towards a circular economy oriented paper recycling industry. *BioResources, 17*(2), 3178–3201.

Maiurova, A., Kurniawan, T. A., Kustikova, M., Bykovskaia, E., Othman, M. H. D., Singh, D., & Goh, H. H. (2022). Promoting digital transformation in waste collection service and waste recycling in Moscow (Russia): Applying a circular economy paradigm to mitigate climate change impacts on the environment. *Journal of Cleaner Production, 354*.

Maryev, V. A., & Smirnova, T. S. (2021). Perspective of CE in Russia: From theory to practice. In *Circular economy: Recent trends in global perspective* (pp. 259–279). https://doi.org/10.1007/978-981-16-0913-8_8

Mativenga, P. T., Agwa-Ejon, J., Mbohwa, C., Sultan, A. A. M., & Shuaib, N. A. (2017). Circular economy ownership models: A view from South Africa industry. *Procedia Manufacturing, 8*, 284–291.

Miranda, I. T. P., Fidelis, R., Fidelis, D. A. S., Pilatti, L. A., & Picinin, C. T. (2020). The integration of recycling cooperatives in the formal management of municipal solid waste as a strategy for the circular economy The case of Londrina. *Brazil. Sustainability (Switzerland), 12*(24), 1–22.

Molla, A. H., Shams, H., Harun, Z., Ab Rahman, M. N., & Hishamuddin, H. (2022). An assessment of drivers and barriers to implementation of circular economy in the end-of-life vehicle recycling sector in India. *Sustainability (Switzerland), 14*(20).

Ohiomah, I., & Sukdeo, N. (2022). Challenges of the South African economy to transition to a circular economy: a case of remanufacturing. *Journal of Remanufacturing, 12*(2), 213–225.

Olatayo, K. I., Mativenga, P. T., & Marnewick, A. L. (2021). Comprehensive evaluation of plastic flows and stocks in South Africa. *Resources, Conservation and Recycling, 170.*

Ottoni, M., Dias, P., & Xavier, L. H. (2020). A circular approach to the e-waste valorization through urban mining in Rio de Janeiro, Brazil. *Journal of Cleaner Production, 261.* https://doi.org/10.1016/j.jclepro.2020.120990

Owojori, O. M., Mulaudzi, R., & Edokpayi, J. N. (2022). Student's Knowledge, Attitude, and Perception (KAP) to solid waste management: A survey towards a more circular economy from a rural-based tertiary institution in South Africa. *Sustainability (Switzerland), 14*(3).

Page, M. J., McKenzie, J. E., Bossuyt, P. M., Boutron, I., Hoffmann, T., Mulrow, C. D., ... & Moher, D. (2021). The PRISMA 2020 Statement: An Updated Guideline for Reporting Systematic Reviews. *BMJ*, n71. https://doi.org/10.1136/bmj.n71

Saha, I., Smirnova, T. S., & Maryev, V. A. (2021). Implementation of eco-industrial park for effectual establishment of circular economy in Russia. *Nature Environment and Pollution Technology, 20*(5), 2031–2040.

Sehnem, S., Ndubisi, N. O., Preschlak, D., Bernardy, R. J., & Santos Junior, S. (2020). Circular economy in the wine chain production: Maturity, challenges, and lessons from an emerging economy perspective. *Production Planning and Control, 31*(11–12), 1014–1034.

Siddiqui, A., & Pandit, R. K. (2021). Smart cities in India: Linkages with circular economy. In Proceedings of *Environmental Science and Engineering*, pp. 185–200. https://www.scopus.com/inward/record.uri?eid=2-s2.0-85106203541&doi=10.1007%2f978-3-030-61891-9_12&partnerID=40&md5=49c208527f40bcc6fdc4a5cdd4ba67d2

Sohal, A., Nand, A. A., Goyal, P., & Bhattacharya, A. (2022). Developing a circular economy: An examination of SME's role in India. *Journal of Business Research, 142*, 435–447. https://doi.org/10.1016/j.jbusres.2021.12.072

Steuer, B. (2021) Identifying effective institutions for China's circular economy: bottom-up evidence from waste management. *Waste Management Research, 39*(7), 937–946.https://doi.org/10.1177/0734242X20972796

Tang, J., Tong, M., Sun, Y., Du, J., & Liu, N. (2020) A spatio-temporal perspective of China's industrial circular economy development. *Science of the Total Environment, 706*, 135754. https://www.sciencedirect.com/science/article/abs/pii/S0048969719357493

van der Velden, R., da Fonseca-Zang, W., Zang, J., Clyde-Smith, D., Leandro, W. M., Parikh, P., Borrion. A., & Campos, L. C. (2022) Closed-loop organic waste management systems for family farmers in Brazil. *Environmental Technology, 43*(15), 2252–2269. https://doi.org/10.1080/09593330.2021.1871660

Vasilenok, V. L., Ignatieva, T. A., Liubarskaia, M. A., & Pilyavsky, V. P. (2021). Circular economy and digital technologies as a tool for overcoming the post-pandemic crisis in Russia. In D. B. Solovev, V. V. Savaley, A. T. Bekker, & V. I. Petukhov (Eds.), *Proceeding of the International Science and Technology Conference "FarEastCon 2020"*. Smart Innovation, Systems and Technologies (Vol. 227, pp. 1135–1141). Springer. https://doi.org/10.1007/978-981-16-0953-4_107

van Zyl, A., & Jooste, J. L, (2022) Retaining and recycling water to address water scarcity in the City of Cape Town. *Development Southern Africa, 39*(2), 108–125. https://doi.org/10.1080/0376835X.2020.1801387

World Bank (World Development Indicators, 2023). *World development indicators*. https://databank.worldbank.org/source/world-development-indicators

Zeng, X., & Li, J. (2020) Circular economy in China. In S. K. Ghosh (Ed.) *Circular economy: Global perspective* (pp. 123–129). Springer. https://link.springer.com/chapter/10.1007/978-981-15-1052-6_6

Zhang, A., Venkatesh, V. G., Wang, J. X., Mani, V., Wan, M., & Qu, T. (2021) Drivers of industry 4.0-enabled smart waste management in supply chain operations: a circular economy perspective in China. *Production Planning & Control, 34*(10). https://www.tandfonline.com/doi/abs/10.1080/09537287.2021.1980909

Zhang, A., Xie, S., Gong, Y., Li, C., & Liu Y (2023) Barriers to compulsory waste sorting for a circular economy in China. *Journal of Environmental Management, 342*, 118180. https://www.sciencedirect.com/science/article/abs/pii/S0301479723009684

Appendix 1: Selected Studies on Circular Economy in Brazil

N. Tshikovhi et al. (eds.), *The BRICS Imperative*,
https://doi.org/10.1007/978-3-032-05932-1

Authors and Year	*Circular Economy Practices*	*Challenges*	*Opportunities*	*Policy Implications*	*Environmental Impact*	*Economic Performance*	*Social and Equity Considerations*	*Technology and Innovation*	*Knowledge Sharing and Capacity Building*
Brazil									
Sehnem S.; Ndubisi N.O.; Preschlak D.; Bernardy R.J.; Santos Junior S. (2020)	Circular economy practices in the wine chain production, specifically the Santa Catarina Association of Fine Wine Producers of Height (ACAVITIS)	Optimizing established circular economy practices for more efficiency	Relation between circular economy and business model maturity	N/A	N/A	Efficiency can impact positively in the family wine production model	Circular economy practices are already established but need optimization	N/A	Suggestion for new studies to confirm the relationship between circular economy practices and maturity stages of business models
Coelho S.T.; Diaz-Chavez R.; Cortez C.L.; Perecin D.; Possetti G.R.C.; Rietow J.C.; Silva C. (2021)	Circular Economy in Brazil with focus on bioenergy and bioeconomy sectors	Inequalities in urban areas and early stages of Circular Economy in Brazil	Improvement of basic sanitation through sustainable use of residues	Importance of legislation and enforcement which addresses recycling, reverse logistic, and energy conversion of urban residues	Sustainable use of waste	N/A	Significant inequalities in urban areas	Bioenergy programs, energy conversion from biomass residues	Overview of the Circular Economy in Brazil and presentation of recent policies and programs

(continued)

(continued)

Authors and Year	*Circular Economy Practices*	*Challenges*	*Opportunities*	*Policy Implications*	*Environmental Impact*	*Economic Performance*	*Social and Equity Considerations*	*Technology and Innovation*	*Knowledge Sharing and Capacity Building*
Guzzo D.; Rodrigues V.P.; Pigosso D.C.A.; Mascarenhas J. (2022)	Brazilian Industrial Agreement for Electrical and Electronic Equipment that governs responsibilities and targets for nationwide collection and treatment of Waste from Electrical and Electronic Equipment (WEEE)	Crucial to analyse the possible effects of public policies, incentives, and infrastructure before implementation	All policy scenarios showed improvement in the EEE collection	Modelling and simulation can inform strategic decision-making in public policies for CE transitions	N/A	N/A	N/A	System dynamics simulation model adapted for the use of smart-phones in Brazil	Investigation of the BIAEEE targets

(continued)

(continued)

Authors and Year	*Circular Economy Practices*	*Challenges*	*Opportunities*	*Policy Implications*	*Environmental Impact*	*Economic Performance*	*Social and Equity Considerations*	*Technology and Innovation*	*Knowledge Sharing and Capacity Building*
de Souza M.; Pereira G.M.; Lopes de Sousa Jabbour A.B.; Chiappetta Jabbour C.J.; Trento L.R.; Borchardt M.; Zvirtes L. (2021)	Marketing strategies underpinned by digital technologies, such as Big Data predictive analytics for reducing food waste in supermarkets	Little knowledge on the role of emerging digital technologies in sustainability of supermarket chains in emerging economies	Actions to reduce negative effects of food waste deterioration	N/A	Reduction of food waste	Benefits both consumers and societal well-being	N/A	Use of sensors and augmented reality to manage dynamic pricing, storage, and item display	Demonstration of how digital technologies can help supermarkets in driving corporate sustainability
Miranda I.T.P.; Fidelis R.; Fidelis D.A.S.; Pilatti L.A.; Picinin C.T. (2020)	Integration of recycling cooperatives in the formal management of municipal solid waste	Challenges in implementing circular economy actions in developing countries	Increase in productivity and professionalization of waste pickers (catadores)	Legal responsibility of cooperatives in the management of MSWRP	Reduction of waste in landfills	Improvement of municipal waste rates	Inclusion of waste pickers in the formal sector of MSWRP management	N/A	Benchmarking option for other cooperatives in relation to their organizational and operational practices
Ottoni M.; Fonseca D.L.; Pertel M. (2022)	Implementation of Waste Management Plans (WMPs) in universities	Only 17% of universities had a WMP; 55% of these were restricted to healthcare services waste	Improve circular and sustainable waste management in universities	Mandatory use of WMPs in universities	Reduction of waste in universities	N/A	N/A	Use of qualitative indicators for analysis	Knowledge sharing through comparison with the international context

(continued)

(continued)

Authors and Year	*Circular Economy Practices*	*Challenges*	*Opportunities*	*Policy Implications*	*Environmental Impact*	*Economic Performance*	*Social and Equity Considerations*	*Technology and Innovation*	*Knowledge Sharing and Capacity Building*
Jesus R.H.G.D.; Barros M.V.; Salvador R.; Souza J.T.D.; Piekarski C.M.; Francisco A.C.D. (2021)	Model to form strategic partnerships for biogas production	N/A	Identifying optimal locations for building biodigesters	N/A	Potential to reduce waste by using it for biogas production	N/A	N/A	Use of agglomerative hierarchical algorithm with multi-criteria analysis and GIS	Application of the model to different regions
Guimaraes Araújo M.; Lovón-Canchumani G.; Bechara Elabras Veiga L. (2021)	Circular Economy evaluation model for lubricating oil re-refining	Impacts of waste collection and transport	Environmental impacts reduction and economic gains	N/A	Reduction of environmental impact	Economic gains from loop closing	N/A	Use of Life Cycle Assessment tool and Simapro software	Creation of a CE evaluation model (ECBR)
Devendran A.A.; Mainali B.; Khatiwada D.; Golzar F.; Mahapatra K.; Toigo C.H. (2023)	Identifying suitable sites for waste transfer stations, route optimization, and landfill lifespan analysis	Challenges with existing MSW management practices	Savings in fuel cost, extension of landfill lifespan	Crucial adoption of the MSW scenario for a circular economy	Reduction of pollution due to landfill leachate	Potential fuel cost savings of 1.5 million BRL/year	N/A	Use of GIS-based spatial analysis	Identification of suitable sites for transfer stations, and route optimization

(continued)

(continued)

Authors and Year	*Circular Economy Practices*	*Challenges*	*Opportunities*	*Policy Implications*	*Environmental Impact*	*Economic Performance*	*Social and Equity Considerations*	*Technology and Innovation*	*Knowledge Sharing and Capacity Building*
van der Velden R.; da Fonseca-Zang W.; Zang J.; Clyde-Smith D.; Leandro W.M.; Parikh P.; Borrion A.; Campos L.C. (2022)	Closed-loop systems for organic waste management	Lack of knowledge about the benefits of recycling organic waste	Diversification of income sources and improvement of agricultural practices	N/A	Management of organic waste, reduction of environmental impact	Generation of revenue from waste	Increase of resilience and income for small-scale farmers	Anaerobic Digestion technology for waste management	Proposal of conceptual designs for closed-loop systems

Appendix 2: Selected Studies on Circular Economy in Russia

N. Tshikovhi et al. (eds.), *The BRICS Imperative*,
https://doi.org/10.1007/978-3-032-05932-1

Authors and Year	*Circular Economy Practices*	*Challenges*	*Opportunities*	*Policy Implications*	*Environmental Impact*	*Economic Performance*	*Social and Equity Considerations*	*Technology and Innovation*	*Knowledge Sharing and Capacity Building*
Russia									
Saha I.; Smirnova T.S.; Maryev V.A. (2021)	Implementation of Eco-Industrial Parks (EIPs) and industrial symbiosis	Developing and implementing an effective EIP model	Effective management of waste and increase in resource efficiency	Policies to support the establishment of EIPs	Reduction of waste disposal, thus, decreasing environmental pollution	Potential for economic growth and job creation	Resolution of regional social problems	Innovative infrastructure and process development for EIPs	The five-stage methodology for EIP organization can be shared for learning and capacity building
Fedotkina O.; Gorbashko E.; Vatolkina N. (2019)	Moving towards circular business models and adopting waste management techniques	Initial stage of circular economy development; defining roles and responsibilities in waste management ecosystem	Opening of new markets, innovative solutions, cost savings	Policy reforms for waste management and allocation of roles	Better waste management, potentially reducing pollution	Possibility of increased savings and access to new markets	Not discussed explicitly	Not discussed explicitly	Case study sharing for best practices in waste management
Kalchenko O.; Evseeva S.; Evseeva O.; Plis K. (2019)	Use of environmentally friendly technologies and principles of circular economy	Needs attention and support from all stakeholders for infrastructure development	International cooperation and integration for better and faster solutions	Policies supporting environmentally-friendly development and international cooperation	Reduction of carbon emissions through energy optimization	Not discussed explicitly	Not discussed explicitly	Adoption of environmentally friendly technologies	Sharing of international experience for better understanding and practice

(continued)

(continued)

Authors and Year	*Circular Economy Practices*	*Challenges*	*Opportunities*	*Policy Implications*	*Environmental Impact*	*Economic Performance*	*Social and Equity Considerations*	*Technology and Innovation*	*Knowledge Sharing and Capacity Building*
Dorokhina E.Yu.; Kharchenko S.G. (2021)	Transition to a circular economy with recycling of production waste into secondary resources	Changing the short-term profit mindset; creating an efficient system for waste collection, sorting, recycling, and disposal	Closed-loop economy; optimization of resource cycles	Policies supporting the use of energy-saving and resource-saving technologies	Decrease in waste accumulation and pollution	Not discussed explicitly	Not discussed explicitly	Implementation of best available technologies	Learning from pilot projects and global garbage reform initiatives
Maryev V.A.; Smirnova T.S. (2021)	Introduction of Extended Producer Responsibility principle, separate waste collection systems, and creation of eco-industrial parks	Lack of systematic approach to waste processing and required institutional base	Involvement of waste into the economy; creation of 70 eco-industrial parks as a solution to waste management	Policies making producers responsible for recycling their goods after use	Reduction in waste generation and increased waste recycling	Economic benefits from cooperation in eco-industrial parks	Not discussed explicitly	Sharing technologies, introduction of best available technologies (BAT)	Not discussed explicitly

(continued)

(continued)

Authors and Year	*Circular Economy Practices*	*Challenges*	*Opportunities*	*Policy Implications*	*Environmental Impact*	*Economic Performance*	*Social and Equity Considerations*	*Technology and Innovation*	*Knowledge Sharing and Capacity Building*
Kostygova L. (2019)	Use of waste from production in Russian industry	Intensive formation of waste, need for economic assessment of waste utilization	Organization of circular economy, rational use of resources	N/A	Nature protection effect from waste utilization	N/A	N/A	N/A	N/A
Vasilenok V.L.; Ignatieva T.A.; Liubarskaia M.A.; Pilyavsky V.P. (2021)	Transition to circular economy in waste management	Low recycling rate, lack of material interest	Digital technologies for waste collection, delivery, and processing	Implementation of extended producer responsibility principle	N/A	Contribution of waste management to GDP is low	N/A	Use of "smart" technologies	N/A
Maiurova A.; Kurniawan T.A.; Kustikova M.; Bykovskaia E.; Othman M.H.D.; Singh D.; Goh H.H. (2022)	Digital transformation in waste collection and recycling	Unorganized waste collection and recycling system	Digitalization to enhance waste recycling industry	N/A	Reduction in GHG emissions through digital technology	Lower cost of operating trucks up to 75%	Job creation in the informal waste sector	Convolutional neural network–based identification system, robotic deployment	N/A

(continued)

(continued)

Authors and Year	*Circular Economy Practices*	*Challenges*	*Opportunities*	*Policy Implications*	*Environmental Impact*	*Economic Performance*	*Social and Equity Considerations*	*Technology and Innovation*	*Knowledge Sharing and Capacity Building*
Abu-Qdais H.A.; Kurbatova A.I. (2022)	Promotion of circular economy through eco-industrial parks	Institutional, technical, and social barriers	N/A	N/A	Reduction in environmental impacts including global warming potential, eutrophication, acidification, smog, and ozone depletion	Circular economy market value in Russia is USD$ 755.05 billion	N/A	N/A	Russian higher educational institutions are key in creating critical mass of experts
Gadzhiev N.G.; Murzak N.A.; Mitenkova A.E.; Skripkina O.V.; Konovalenko S.A. (2020)	Formation of circular economy in Russia for sustainable development	Barriers to implementing business models of circular economy	Transition to renewable energy	N/A	Reduction in greenhouse gas emissions	N/A	Balance between environmental, economic and social components of sustainable development	N/A	N/A

Appendix 3: Selected Studies on Circular Economy in India

N. Tshikovhi et al. (eds.), *The BRICS Imperative*,
https://doi.org/10.1007/978-3-032-05932-1

Authors and Year	*Circular Economy Practices*	*Challenges*	*Opportunities*	*Policy Implications*	*Environmental Impact*	*Economic Performance*	*Social and Equity Considerations*	*Technology and Innovation*	*Knowledge Sharing and Capacity Building*
India									
Kapoor N.; Jauhari S.; Maheshwari D. (2022)	Focused on three paradigms: reduce, reuse, and recycle via business models	Not clearly mentioned	Lessons that India can share with others through the analysis of business models	Not clearly mentioned	N/A but refers to system restorativeness in industry processes	Explored via case study of El Rhino's business model	Not mentioned	Business models analysed for take-make-reduce and take-make-recycle systems	Shared knowledge via exploration of multiple business models
Debnath B.; Das A.; Das A. (2021)	Urban mining and resource recovery from e-waste	Informal sector handling the majority of e-waste. Registered e-waste recyclers cannot operate at full capacity	Explored possibilities of CE in e-waste sector. Unity in diversity of India allows for bringing together stakeholders for CE	Emphasized extended producer responsibility and CE principles in national policies	E-waste as a huge source of secondary raw materials	Not mentioned	Diverse stakeholders from allied industries can collaborate	E-waste sector itself is a platform for technological solutions	N/A

(continued)

(continued)

Authors and Year	*Circular Economy Practices*	*Challenges*	*Opportunities*	*Policy Implications*	*Environmental Impact*	*Economic Performance*	*Social and Equity Considerations*	*Technology and Innovation*	*Knowledge Sharing and Capacity Building*
Siddiqui A.; Pandit R.K. (2021)	Smart practices towards regenerative system: waste management, e-governance, and smart mobility	Challenges from design to implementation phase of smart city projects	The integration of CE principles in smart city objectives	Required policy-level reforms and institutional capacity building	N/A	Increase in economic resilience of the city and its citizens	Advocated for uplifting infrastructure, and introducing financing mechanisms	Smart city practices as an innovation platform	Local traditions integrating the 6Rs and smart city practices provide opportunities for capacity building
Molla A.H.; Shams H.; Harun Z.; Ab Rahman M.N.; Hishamuddin H. (2022)	Substitution of end-of-life concept with restoration in the ELV recycling sector	Limited technology, financial constraints, and lack of knowledge and expertise	Economic viability, environmental degradation, and global agenda as primary drivers	Importance of appropriate policies, strategies, and a conducive legal framework	Environmental degradation as a driver for CE implementation	Economic viability as a primary driver	N/A	Limited technology as a significant barrier	Lack of knowledge and expertise as a significant barrier

(continued)

(continued)

Authors and Year	*Circular Economy Practices*	*Challenges*	*Opportunities*	*Policy Implications*	*Environmental Impact*	*Economic Performance*	*Social and Equity Considerations*	*Technology and Innovation*	*Knowledge Sharing and Capacity Building*
Singh S.; Dasgupta M.S.; Routroy S. (2022)	Identified and categorized 23 CSFs in six implication dimensions for e-waste collection policy	Not clearly mentioned	Not clearly mentioned	Focused on design of e-waste collection policy and importance of CSFs	N/A	Business sustainability to stakeholders	Stakeholders include a wide range from manufacturers to regulators	Technology involvement as a prominent influencing CSF	N/A
Hossain et al. (2022)	Focused on plastic waste management, addressing reverse supply chain, effective PW management, source-specific recovery, and PW rules	A lack of understanding of the depth of the plastic waste problem and potential solutions	Identifying implementable strategies for policymakers and research opportunities for future researchers	Need for effective policies for holistic PW management and recycling	Not specified	Not specified	Not specified	Use of technology in managing plastic waste	Addresses the knowledge gap in understanding the depth of the problem and possible solutions

(continued)

(continued)

Authors and Year	*Circular Economy Practices*	*Challenges*	*Opportunities*	*Policy Implications*	*Environmental Impact*	*Economic Performance*	*Social and Equity Considerations*	*Technology and Innovation*	*Knowledge Sharing and Capacity Building*
Ahmed et al. (2023)	Resource value extension based circular economy model in the Paperman Foundation	Limited research on cases based in emerging economies	Achievement of lowered carbon emissions in emerging and transitional economies	Not specified	Lowered carbon emissions	Not specified	Not specified	Not specified	Calls for more case studies in emerging economies
Ghosh et al. (2021)	Focuses on the reduction, reuse, and recycling of waste through policy frameworks and business models	Untreated sewage and pollution from untreated waste	Waste management becoming business models through recycling and recovery techniques	Emphasis on the Environment Protection Act and other rules supporting sustainable waste management	Pollution reduction from all potential sources	Robust economic performance from the effective management of waste	Not specified	Use of various waste utilization technologies	Not specified

(continued)

(continued)

Authors and Year	*Circular Economy Practices*	*Challenges*	*Opportunities*	*Policy Implications*	*Environmental Impact*	*Economic Performance*	*Social and Equity Considerations*	*Technology and Innovation*	*Knowledge Sharing and Capacity Building*
Kamble et al. (2021)	Circular economy model for waste management	Not specified	Not specified	Not specified	Not specified	Not specified	Not specified	Utilization of the Technique for Order Preference by Similarity to Ideal Solution (TOPSIS) SIMOS approach	Identification of the critical success factors for waste management
Sohal et al. (2022)	The establishment of a circular economy in SMEs in India based on a culture that focuses on less waste, more recycling, reusing, repurposing, and repairing of parts and products	Limited support for SMEs through government improvement programs	Not specified	Recommendations for managers, policy-makers, and industry associations to help SMEs transition to a circular economy	Not specified	Not specified	Not specified	Application of socio-technical system (STS) theory	Micro-level insights from STS theory to identify various enablers and drivers

Appendix 4: Selected Studies on Circular Economy in China

N. Tshikovhi et al. (eds.), *The BRICS Imperative*,
https://doi.org/10.1007/978-3-032-05932-1

Authors & Year	*Circular Economy Practices*	*Challenges*	*Opportunities*	*Policy Implications*	*Environmental Impact*	*Economic Performance*	*Social and Equity Considerations*	*Technology and Innovation*	*Knowledge Sharing and Capacity Building*
China									
Zhang et al. (2023)	Compulsory waste sorting for a circular economy	Hasty and inappropriate planning, lack of policy support at grassroots level	Effective waste management	Policy support required to implement effective waste sorting	Reduction in waste sent to landfills	Not mentioned	Not mentioned	Not mentioned	Highlighted barrier studies
Zeng et al. (2019)	Improvement of resource efficiency and recycling rate	Regional resource shortage, supply chain disruption	Cost decrease, technology innovation	Regulatory control required to enforce circular economy practices	Reduces waste and optimizes resource use	Not mentioned	Not mentioned	Technology innovation as a driver of circular economy	Not mentioned
Kurniawan et al. (2022)	Transformation of solid waste management through digitalization	Over-generation of municipal solid waste	Effective management of municipal solid waste	Not mentioned	Reduces environmental pollution, promotes 3Rs (Reduce, Reuse, Recycle)	Creation of job opportunities	Not mentioned	Digitalization as a driving force	Helps decision-makers in planning a smooth transition to a circular economy

(continued)

(continued)

Authors & Year	*Circular Economy Practices*	*Challenges*	*Opportunities*	*Policy Implications*	*Environmental Impact*	*Economic Performance*	*Social and Equity Considerations*	*Technology and Innovation*	*Knowledge Sharing and Capacity Building*
Li et al. (2022)	Reinvent and rethink, restore, reduce and avoid, recirculation of resources	Not mentioned	Increased financial performance from reinvent and rethink, and restore, reduce and avoid practices	Not mentioned	Reduction in waste generation	Positive impact on financial performance through certain practices	Not mentioned	Digital supply chain platforms	Not mentioned
Wang H. et al. (2020)	Rebuilding economy to circular model, enhancing material use efficiency	Need for datasets, metrics, indicators	Potential for increased circularity; End of life waste recycling improvement	Circular Economy Promotion Law; Policy indicators for Green Development Indicator System, Ecological Civilization Construction Assessment, SDGs	Increasing circularity; Improved end of life waste recycling	Strong physical growth dynamic; Transition to closed-loop systems	Consideration for a consumption-based economy; Equity in transition	Potential for future circular economy; Transition to closed loop systems	Using policy indicators for targets and transition

(continued)

(continued)

Authors & Year	*Circular Economy Practices*	*Challenges*	*Opportunities*	*Policy Implications*	*Environmental Impact*	*Economic Performance*	*Social and Equity Considerations*	*Technology and Innovation*	*Knowledge Sharing and Capacity Building*
Tang et al., 2020	The "3Rs" principle in industrial development	Not optimistic outlook for emissions of major pollutants	Resource output, resource consumption, and resource reuse and waste disposal have improved	Dependence on the regional economic situation for capital input	N/A	Tied to regional economic conditions	N/A	N/A	N/A
Chen et al. (2020)	Pilot cities for circular economy construction	Gradual recovery from initial economic decline in pilot cities	Sustainable economic growth over time	Role of circular economy policy in local economic growth	N/A	GDP growth rate decreases initially but recovers over time	N/A	N/A	N/A
Kuo & Chang (2021)	Firms' responsible production following circular economy legislation	Disclosure of circular economy information to meet the information needs of stakeholders	Higher sustainable growth rate and return on equity for disclosing firms	Role of legislation in guiding firms' responsible production	N/A	Impact of circular economy information disclosure on firms' economic sustainability	Role of state-owned enterprises in leading implementation	N/A	Filling the literature gap related to circular economy actions in developing countries

(continued)

(continued)

Authors & Year	*Circular Economy Practices*	*Challenges*	*Opportunities*	*Policy Implications*	*Environmental Impact*	*Economic Performance*	*Social and Equity Considerations*	*Technology and Innovation*	*Knowledge Sharing and Capacity Building*
Steuer (2021)	Advancement of circular economy in waste management	N/A	Potential innovation of current formal WM systems in urban China	Informal institutional elements in waste collection and pre-processing	N/A	N/A	N/A	N/A	Identification of key institutional ingredients for effective outcomes
Zhang et al. (2021)	Industry 4.0-enabled smart waste management in supply chain operations	Operational challenges in implementing smart waste management	Overcoming operational challenges, recovering value, speeding up operations, saving cost and improving profit	N/A	N/A	Impact on supply chain operations and profit	N/A	Industry 4.0 technologies in waste management	Analysis of key drivers and their interrelationships

Appendix 5: Selected Studies on Circular Economy in South Africa

N. Tshikovhi et al. (eds.), *The BRICS Imperative*,
https://doi.org/10.1007/978-3-032-05932-1

Article	*Circular Economy Practices*	*Challenges*	*Opportunities*	*Policy Implications*	*Environmental Impact*	*Economic Performance*	*Social and Equity Considerations*	*Technology and Innovation*	*Knowledge Sharing and Capacity Building*
South Africa									
Bag et al. (2022)	Eco-innovation, green supply chain management (GSCM) practices, CE capability, big data driven supply chains (BDSC)	Understanding nuances and relationships between various practices and their impacts	Strong positive relationships between institutional pressures, eco-innovation, GSCM, and CE capability	The role of institutional pressures in enabling SMEs	Lower environmental footprint due to green practices	Enhanced performance of SMEs	Social structures of organizations play a key role	The role of big data in supply chains	Frameworks require further validation and explanation
Bag et al. (2021)	Adoption of big data analytics-powered artificial intelligence, sustainable manufacturing practices	Lack of understanding why firms adopt big data analytics-powered AI	Institutional pressures on resources can influence the adoption of AI and sustainable practices	Policies that support the adoption of AI can enhance sustainable manufacturing and circular economy capabilities	Reduction of environmental footprint through AI-powered sustainable manufacturing	Improved supply chain performance	N/A	The role of big data analytics-powered AI	Understanding the interplay of resources, institutional pressures, and AI
Adeleke et al. (2021)	Waste management strategies	Unmanaged municipal solid wastes, inefficiency of current waste management systems	Accurate and reliable waste-related data can fast-track sustainable waste management	Policies are needed to support data-driven waste management decisions	Potential reduction of detrimental effects on human health and the environment	N/A	The impact of waste on human health and the environment	N/A	A data-mining approach can aid waste management decisions and optimize investments

(continued)

(continued)

Article	*Circular Economy Practices*	*Challenges*	*Opportunities*	*Policy Implications*	*Environmental Impact*	*Economic Performance*	*Social and Equity Considerations*	*Technology and Innovation*	*Knowledge Sharing and Capacity Building*
Godfrey et al. (2021)	Repair and reuse of end-of-life products, recycling of grey water, use of renewable energy	N/A	Circular economy provides opportunities to address resource scarcity and security	Circular economy can support South Africa's local and international environmental commitments	Reduction of water and energy crisis	Supports local manufacturing, reduces dependence on resource imports, creates jobs	Addresses high levels of unemployment, poverty, and inequality	N/A	Informal sector plays a key role in unlocking a circular economy
Olatayo et al. (2021)	Plastic material flows, Waste Research Development and Innovation (RDI) Roadmap	Lack of complete and recent data, underlying assumptions on waste evaluation	The transition to a circular economy can be informed by defining plastic material flows	Waste RDI Roadmap to support the transition to a circular economy	Tackling plastic waste can mitigate environmental pollution	N/A	N/A	A new approach for comprehensive evaluation of recycling rate	The first known case of a Sankey diagram for plastic materials in South Africa
Owojori et al., 2022	Investigates student's KAP to solid waste management in a rural-based tertiary institution	Revealed low and inadequate knowledge of students on solid waste management	Encourages participatory environmental programs and economic incentives to motivate students	Suggests institutionalizing environmental education	Focused on solid waste management and its impact on the environment	Not directly addressed	Highlights the role of students as key agents of transformation in society	Not directly addressed	Emphasizes the role of environmental education in transitioning to a circular economy
van Zyl et al. (2022)	Discusses retaining and recycling water in the City of Cape Town	Not directly addressed	Identifies the retention of rainwater, the reuse of greywater, and treatment of wastewater as potential water sources	Advocates for policy interventions related to alternative water sources	Highlights water shortage and the need to manage water resources better	Not directly addressed	Not directly addressed	Demonstrates the use of a system dynamics model to evaluate policy interventions	Not directly addressed

(continued)

(continued)

Article	*Circular Economy Practices*	*Challenges*	*Opportunities*	*Policy Implications*	*Environmental Impact*	*Economic Performance*	*Social and Equity Considerations*	*Technology and Innovation*	*Knowledge Sharing and Capacity Building*
Botha et al. (2022)	Introduces a decision support system (DSS) for decentralised waste management focusing on sustainable SMME development	Cites the lack of infrastructure, skills, and public funding as challenges in waste management	Reveals over 600 unique combinations for waste management in the DSS model	Not directly addressed	Focuses on waste management and its potential impact on the environment	Discusses how the DSS can support SME business development	Not directly addressed	Introduces a novel decision support system for decentralized general waste management	Discusses the development and validation of a decision support system
Mativenga et al. (2017)	Examines South African industry's view of circular economy models, with a focus on composite waste	Not directly addressed	Identifies cost reduction as a key driver for recycling composites	Highlights the need for national models for managing the transition to a circular economy	Suggests the reduction of waste through recycling and reuse of composite materials	Points out that cost reduction can drive recycling and circular economy initiatives	Not directly addressed	Not directly addressed	Advocates for knowledge sharing about appropriate national models and cost reduction opportunities
ohiomah et al. (2022)	Studies the barriers to implementing circular economy in South Africa through remanufacturing	Identifies barriers such as quality concerns, lack of guidelines, lack of government support, and poor customer perception	Suggests remanufacturing as a viable way to implement circular economy in South Africa	Recommends more government support for remanufacturing	Remanufacturing reduces waste and thereby minimizes environmental impact	Discusses economic benefits of remanufacturing including reduced costs and extended product lifecycle	Not directly addressed	Not directly addressed	Not directly addressed

Index

N. Tshikovhi et al. (eds.), *The BRICS Imperative*,
https://doi.org/10.1007/978-3-032-05932-1

Kolibri 360 GmbH
Ferdinand-Jühlke-Straße 7
99095 Erfurt, Deutschland
produktsicherheit@kolibri360.de